BLUE SKIES AND BENCH SPACE

Adventures in Cancer Research

BLUE SKIES AND BENCH SPACE

Adventures in Cancer Research

KATHLEEN WESTON

Cancer Research UK
London Research Institute

CSH PRESS
COLD SPRING HARBOR LABORATORY PRESS
Cold Spring Harbor, New York • www.cshlpress.org

BLUE SKIES AND BENCH SPACE: ADVENTURES IN CANCER RESEARCH

Published by Cold Spring Harbor Laboratory Press
Printed in the United States of America

Publisher and Acquisition Editor	John Inglis
Director of Editorial Development	Jan Argentine
Project Manager	Inez Sialiano
Permissions Coordinator	Carol Brown
Production Editor	Rena Springer
Desktop Editor	Techset Ltd.
Production Manager	Denise Weiss
Cover Designer	Mike Albano

Front cover: Cover image by Joe Brock (MRC National Institute for Medical Research); merged histology image is from an oblique transverse section through an E14.5 mouse embryo, courtesy of Tim Mohun (MRC National Institute for Medical Research).

Library of Congress Cataloging-in-Publication Data

Weston, Kathleen M., 1961-
Blue skies and bench space : adventures in cancer research / Kathleen M. Weston, Cancer Research UK, London Research Institute.
pages cm
Includes index.
ISBN 978-1-62182-077-2 (hardcover: alk. paper)
1. Cancer--Research--History. 2. Cancer--Research--International cooperation. 3. Medical sciences--International cooperation. 4. Oncologists--Anecdotes. 5. Imperial Cancer Research Fund (Great Britain) I. Title.

RC267.W46 2014
362.19699'4--dc23

2013020491

10 9 8 7 6 5 4 3 2 1

All World Wide Web addresses are accurate to the best of our knowledge at the time of printing.

For a complete catalog of all Cold Spring Harbor Laboratory Press publications, visit our website at www.cshlpress.org.

For Rosa and James

Contents

Preface

In 2016, the Francis Crick Institute, an ambitious multidisciplinary research centre a stone's throw from the British Library and the Eurostar terminal at St. Pancras, is scheduled to open its doors. The Crick's imminent completion undoubtedly signals the beginning of a new era in British science, but for many of its first occupants, the excitement that they feel will be tempered by sadness: two venerable and distinguished London scientific institutions, the Medical Research Council National Institute for Medical Research and the Cancer Research UK London Research Institute, will close upon the opening of the new building, and their inmates will leave behind their much-loved old laboratories for a shared future at the Crick.

This book arose from the determination of the Director of the London Research Institute, Richard Treisman, that his laboratory should not just disappear into the past without a fitting memorial. Some truly great research has happened there—in its past guise as the Imperial Cancer Research Fund and in its present incarnation as the London Research Institute, the laboratory played a central role in the revolution that swept through the biological sciences following James Watson and Francis Crick's 1953 discovery of the structure of DNA, the molecule of life. However, rather than a conventional history, the project has evolved into something a little different. I have chosen eight research stories that encapsulate the scientific adventures of the laboratory's inmates and written about them in what I hope is an entertaining, informative, and determinedly irreverent romp through the highs and lows of laboratory life. I wanted to give a flavour of what it is really like working in a laboratory and also to show that there is no scientific stereotype: The characters in this book come from all walks of life and have frequently followed wildly different paths towards their scientific goals.

If you are in search of a heavy-weight history of molecular biology, this is not the book for you. It is undoubtedly somewhat biased because I have limited my descriptions of past research to that which must be understood in order to put the discoveries I discuss in their proper context. Instead of learned footnotes, I have made some fairly eclectic suggestions for further reading and recommended some useful web links to animations of some of the processes I describe. For those wishing to explore the field in greater depth, the excellent textbook *Molecular Biology of the Cell,* by Bruce Alberts et al., and the website of the Cold Spring Harbor DNA Learning Center (www.dnalc.org) should sate even the most voracious appetites.

So who is this book for? If you are an undergraduate thinking of a life in science or someone just starting a career at the bench, it will give you some idea of how we got to where we are today and what pleasures and pains you can expect in your daily existence in the future. For those already hard at work at the coalface or looking back on a life in science, you'll find much that is familiar, and perhaps a few surprises. If you're an interested spectator willing to learn a little bit of jargon (and I've included a glossary to interpret the worst bits), then I hope very much that this is the book you've been looking for: a rough guide to the slightly peculiar inhabitants, customs, and folk tales of a country that really isn't a lot different from your own (and which, these days, I should add, contains rather more, although still not enough, women).

Finally, a few words regarding the title of the book. Clearly, the skittishly unreliable climate in London is not the reason for calling it *Blue Skies and Bench Space.* Rather, I am very keen to hammer home the notion, becoming increasingly more foreign to some governments and funding organisations, that the key to understanding the biology of life, in sickness and in health, lies in so-called blue skies research—working on projects because they are interesting and answer fundamental questions. Every single story in this book has led to medical advances, some of them immense, yet none of them began with that aim. I can think of no better way to sum this up than to quote J. Michael Bishop, my postdoctoral mentor and the 1989 Nobel Laureate in Physiology or Medicine: "The proper conduct of science lies in the pursuit of nature's puzzles, wherever they may lead. We cannot prejudge the utility of any scholarship; we can only ask that it be sound. We cannot always assault the great problems of biology at will. We must remain alert to nature's clues and seize on them whenever and wherever they may appear."

KATHLEEN WESTON

Alberts B, Johnson A, Lewis J, Raff M, Roberts K, Walter P. 2007. *Molecular biology of the cell,* 5th ed. Garland Science, New York, NY.

Bishop JM. 2003. *How to win the Nobel Prize.* Harvard University Press, Cambridge, MA.

Acknowledgments

I am indebted to the following people for sparing the time to talk to me, sometimes at great length, who in many cases have also read and criticised what I have written about them (needless to say, any factual errors that have crept into the text are entirely my own fault): Adrian Hayday, Andy McMahon, Bob Kamen, Brenda Marriott, Catrin Pritchard, Cliff Tabin, Dave Hancock, David Ish-Horowicz, David Lane, David Page, Frank Fitzjohn, Gerard Evan, Harold Varmus, Iain Hagan, Jacky Hayles, Jim Smith, John Cairns, John Diffley, Julian Downward, Lionel Crawford, Lizzy Fisher, Melanie Lee, Mike Fried, Mike Owen, Mike Waterfield, Paul Nurse, Peter Goodfellow, Peter Parker, Phil Ingham, Richard Treisman, Robin Lovell-Badge, Sergio Moreno, Sheena Pinchin, Steve West, Tim Hunt, Tomas Lindahl, Tony Carr, Trevor Littlewood, and Walter Bodmer. In addition, I thank Birgit Lane, Connie Casey, Doug Green, Ed Harlow, Errol Friedberg, Mike Billmore, Nissim Hay, Peter Newmark, and Tony Hunter for some really helpful e-mail exchanges.

Some of the photographs in this book have been acquired by pestering the following people to ransack their photo albums, for which I am very grateful: Andy McMahon, Anne Nurse, Ashley Dunn, Jim Smith, Lionel Crawford, Martin McMahon, Melanie Lee, Peter Goodfellow, Peter Parker, Phil Ingham, Robin Lovell-Badge, Sergio Moreno, Steve West, Tim Hunt, Trevor Littlewood, and Viesturs Simanis. Thanks to David Bacon of the LRI for help with electronic conversions.

Finally, I need to say a heartfelt thank you to the people who have read my inept drafts, commented on them without making me cry, and/or generally been a huge support throughout this project: Jim Smith, Ian Armitage, Kwĕsi Edman and the Shropshire Cellists, Richard Treisman and Ava Yeo of the London Research Institute, and John Inglis and Alex Gann at Cold Spring Harbor Laboratory Press.

CHAPTER 1

Beginnings

The Dawn of Molecular Biology: What Is Life?

In 1944, the theoretical physicist and Nobel Laureate Erwin Schrödinger published *What Is Life?*, based on a series of public lectures he had given in Dublin, where he had settled after fleeing Austria following the arrival of the Nazis in 1938. The little book, less than 100 pages long, is a unique mixture of quantum physics, the biology of heredity, and an unexpected dose of mysticism, and became a bestseller, with eventual sales topping 100,000 copies. In it, Schrödinger publicised to an English-speaking audience a paper written in an obscure German journal in 1935 by the theoretical physicist Max Delbrück and his colleagues Nikolay Timoféeff-Ressovsky and Karl Zimmer. The "Three Man Paper," as it came to be known, suggested that, just as for inorganic matter, life itself had to be governed by the laws of physics. From this it followed that genes, rather than being just the geneticists' theoretical units of inheritance, might actually be real things that could be studied at the molecular level. In addition to enthusiastically espousing this hitherto esoteric view, Schrödinger dangled in front of his readers the tantalising possibility that exploring biology might reveal new laws of physics: "it emerges that living matter, while not eluding the 'laws of physics' as established up to date, is likely to involve 'other laws of physics,' hitherto unknown, which, however, once they are revealed, will form just as integral a part of this science as the former" (Schrödinger 1944).

Schrödinger's book, combined with a surprising romanticism in his readership, was responsible for recruiting an outstanding cohort of physicists to populate the new field of molecular biology; it should be noted, however, that he misconstrued some of the 1935 paper's ideas, and much later on, the crystallographer Max Perutz, whose own conversion had preceded Schrödinger's, observed that "what was true in his book was not original, and most of what was original was known not to be true even when it was written" (Perutz 1987).

Looking back with a modern eye on the rumpus that *What Is Life?* caused in the physics world in the 1940s, it is hard to understand how the simple statements that biology and physics had things in common and that there were experimental techniques that could be applied to finding out exactly what genes were could have been such novel concepts. Probably, much of the secret of the book's success lay in its timeliness. In François Jacob's words:

> After the war, many young physicists were disgusted by the military use that had been made of atomic energy. Moreover, some of them had wearied of the turn experimental physics had taken … of the complexity imposed by the use of big machines. They saw in it the end of a science and looked around for other activities. Some looked to biology with a mixture of diffidence and hope. Diffidence because they had about living beings only the vague notions of the zoology and botany they remembered from school. Hope, because the most famous of their elders had painted biology as full of promise. To hear one of the fathers of quantum mechanics ask himself: "What Is Life?" and to describe heredity in terms of molecular structure, of interatomic bonds, of thermodynamic stability, sufficed to draw towards biology the enthusiasm of young physicists and to confer on them a certain legitimacy (Jacob 1970).

Fortunately for Schrödinger's readers, the new field he had described so persuasively was already up and running; the source of the book's inspiration, Max Delbrück, had been rather busy in the decade since the "Three Man Paper" had appeared. Delbrück had been awarded a fellowship in 1937 from the Rockefeller Foundation to leave Germany and learn some fruit fly genetics at the California Institute of Technology (Caltech) in Pasadena. Luckely for molecular biology, the *Drosophila* genetics literature was by that time so convoluted and stuffed with jargon that Delbrück had difficulty making head or tail of it, and instead fell in with Emory Ellis, a physical chemist at Caltech. Ellis was interested in cancer and had reasoned that because it was known that viruses sometimes cause cancer, he should try to work out how. He decided that the best way to do this was to study the simplest possible viruses, those that infect and kill bacteria. To this end, he had taught himself some basic microbiology, built some equipment, and learned how to grow cultures of *Escherichia coli* bacteria on the surface of agar plates, creating a smooth opaque lawn of bacteria atop the jelly-like clear agar after overnight incubation at 37°C. Next, he had collected a bucket of Los Angeles sewage and found in it a bacteriophage, a bacterial virus able to kill *E. coli*; when he overlaid his opaque bacterial lawn with a solution of minute particles purified from the sewage, in the morning he could see clear holes or plaques in the lawn, each showing where a virus particle had infected and killed the surrounding bacteria. By counting the number of holes, one could precisely quantitate the number of infecting viruses.

This plaque assay completely blew Delbrück away: "I was absolutely overwhelmed that there were such very simple procedures with which you could visualize individual virus particles; ...you could put them on a plate with a lawn of bacteria, and the next morning every virus particle would have eaten a 1 mm hole in the lawn. You could hold up the plate and count the plaques. This seemed to me just beyond my wildest dreams of doing simple experiments on something like atoms in biology" (Delbrück interview, Caltech Oral History Archive, with permission from the California Institute of Technology © 1979).

Delbrück took to phage work with a vengeance, although sadly, Ellis dropped out shortly afterwards (his funders, not recognising his prescience, were unable to understand what relevance his work had to cancer research and forced him back into dull mainstream studies transplanting tumours between mice). After the outbreak of war, Delbrück remained in the United States, moving to Vanderbilt University, and in late 1940, he met Salvador Luria, an Italian microbiologist, and Al Hershey, a bacteriologist. The three men applied a combination of genetics, bacteriology, and the rigorous experimental methodology of physics to the task of understanding how phages transmitted their genetic information down the generations and how the information remained stable.

Quite early on, Delbrück realised that to succeed, the phage field needed to accrue disciples as rapidly as possible, and to this end, he and Luria started a summer school in 1945 at the dilapidated but beautiful Cold Spring Harbor Laboratories on Long Island. Aspiring phage researchers could learn the techniques necessary to study phage, have a whale of a time at the multitude of social events, and fall under Delbrück's spell; a charismatic and brilliant man, he inspired great affection and loyalty, despite being alarmingly difficult to please when it came to matters of science. The summer school's aim, that participants would leave as fully indoctrinated phage group members, or at least phage-literate converts, was extremely successful; the phage group's

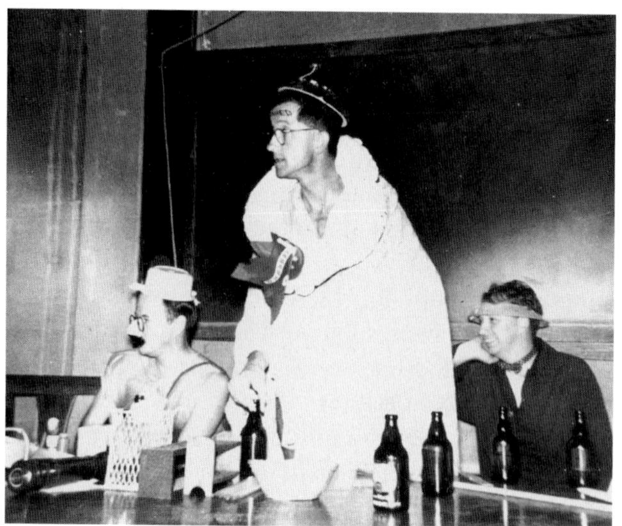

Max Delbrück "on trial" at the 1950 Phage Meeting.
(Photograph courtesy of Cold Spring Harbor Laboratory Archives.)

ethos (with respect to both experimental methodology and how to have a good time) was widely disseminated, to the extent that virtually all molecular biology bloodlines in the United States can ultimately be traced back to Delbrück, Luria, or Hershey. For their own valuable discoveries, and also for their achievement in laying the philosophical foundations of modern molecular biology, the three men were awarded the 1969 Nobel Prize in Physiology or Medicine.

The new molecular biology field was transformed from an intriguing minority discipline into a major scientific force by the arrival in 1951 at the Cavendish Laboratories in Cambridge of a 23-year-old American, Jim Watson. Watson, who had grown up in the phage group as a student with Salvador Luria, had come to Europe to try to learn some biochemistry in Copenhagen but had become so bored that he decamped to England, where he annoyed, entertained, and stimulated his bemused new hosts in approximately equal measure. Watson, true to his phage roots, was obsessed with how hereditary information was transmitted down the generations. DNA had been known for some time to be the hereditary material, thanks to an experiment performed in 1944 by Oswald Avery, Colin MacLeod, and Maclyn McCarty, but nobody knew how it replicated itself or how its information was decoded. Watson realised that the key to understanding how DNA worked was to determine its structure, and this he and Francis Crick did, building their famous double-helical model in a brilliant extrapolation from Rosalind Franklin's beautiful X-ray data.

Watson and Crick's paper "Molecular Structure of Nucleic Acids; A Structure for Deoxyribose Nucleic Acid," published in *Nature* in April 1953, opened up an enormous landscape of biological possibility. In the years that followed, a small and extraordinarily talented band of pioneers showed how DNA was replicated, that DNA made RNA, that messenger RNA (mRNA) made protein, and that the linear sequence of the amino acids comprising a protein was determined by the triplet genetic code that was carried in the DNA, copied (transcribed) into mRNA, and translated into protein on ribosomes. Much to the annoyance of some classical biologists, whose belief in the complexity of nature was almost mystical, and some of the phage group physicists, who were still hoping to fulfil Schrödinger's prediction that they would find new laws of physics by studying biology, the basis of life really was as simple as Francis Crick's central dogma of "DNA makes RNA makes protein" suggested. Molecular biology, with its triumphantly vindicated reductionist outlook, was on a high.

Although Gunther Stent, one of Delbrück's original phage group at Caltech, dismissed the era following the remarkable conceptual leaps leading to the central dogma as "what remained now was to iron out the details" (Stent 1968), to those working in the steadily expanding field it was anything but mundane. Building on the central dogma, molecular biologists turned their attention to filling in the vast information voids that still existed in their new submicroscopic landscape, particularly in terms of how processes were regulated. How did a cell know when to replicate its DNA, to make RNA, and to make protein? How did proteins know where to go in a cell? There were so many truly important questions to be answered, and advances in methodology and technology meant that quite often, the only limit on what could be achieved was the intellectual capacity of the researcher involved.

From Phage to Papovaviruses: Tackling Cancer

Even before the mechanisms governing the prokaryotic bacterial and phage world were partially understood, or Jim Watson had set foot in Cambridge, attention had turned towards the universe of the eukaryota.

Eukaryotes, defined as organisms whose cells contain a nucleus, as opposed to the nonnucleated prokaryotes, range from the simple, such as the unicellular yeasts, to complex multicellular organisms such as ourselves. Understanding the molecular engineering involved in building and maintaining a eukaryote, and importantly, the molecular failures that lead to defects and disease, began to appear possible. Prospective pioneers started to look for ways in which to attack the problem but first had to figure out how to do the kind of experiments that had been the key to the work in prokaryotes. They were stumped by two big problems: the difficulty of getting live cells out of an animal and growing them in the lab and the lack of a tool akin to a bacteriophage with which to probe cell function.

In 1950, a fortuitous collision between extreme wealth and extreme discomfort jump-started the field of eukaryotic molecular biology by fixing both these issues. Max Delbrück, who was back once more in Pasadena as head of a small but flourishing phage lab, received an interesting proposition from Lee DuBridge, President of Caltech. The richest cotton baron in California, Colonel James G. Boswell, had been hospitalised with a serious case of shingles, and finding out from his doctor that almost nothing was known regarding the shingles virus, *Varicella zoster*, which was causing him so much pain, had decided to throw some money at the problem; he was offering Caltech, with its world-famous expertise in virology, $225,000 to spend on research into animal viruses. Boswell's hoped-for cure did not arrive in time to fix his shingles (in the end, his doctor prescribed a daily dose of bourbon, which the formerly teetotal Boswell took to with gusto), but his money was well spent. Delbrück was happy to take on the challenge and decided that because he knew nothing about animal viruses, the most sensible way to proceed would be to spend the first chunk of money from the James G. Boswell Foundation on a conference, getting together the leading lights in the virus world. A motley assembly of 35 plant, animal, and bacterial virologists showed up at Caltech in March 1950, but the conference was not particularly successful; a camping trip to Death Valley immediately afterwards seems to have been the highlight of the meeting, and some useful scientific collaborations were forged in the dust and heat of the desert. The problem was that as far as molecular biology was concerned, plant and animal virology was very much in the Dark Ages; apart from the structural biologists, who were starting to explore how the viruses looked and what they were made of, nobody had even managed to develop a simple way of counting their viruses, let alone doing anything else to determine how they replicated.

After seeing off his new colleagues, Delbrück returned to his lab and decided that if anything was to be done about dragging eukaryotic virology into the modern age, he would have to do it himself. He summoned two of his laboratory members, Renato Dulbecco and Seymour Benzer, and asked whether either of them would be interested in taking on the Boswell Foundation project. Benzer was quite happy with what he was doing and said no. Dulbecco, however, was interested. Originally trained as a medic and therefore with an interest in human disease, he was a bit fed up with his current phage work and agreed to have a look at what could be done in the new system. After a three-month road trip to look in detail at what was going on in the existing animal virus laboratories, he returned to Caltech, where he was promptly banished to a small room in the second basement for his decision to work on the human pathogen Western equine encephalitis virus, which scared the pants off the rest of the department.

Dulbecco realised that to do any meaningful work with animal viruses, he first had to sort out the issue of how to quantitate them, just as bacteriophages could be quantitated in a plaque assay. To set up an animal-cell version of the plaque assay, he needed to find a way of growing viruses on flat lawns of animal cells in tissue culture, rather than in the animals themselves, as was currently done. Fortunately, back in his native Italy, Dulbecco had worked with Giuseppe Levi (mentor of three future Nobel laureates—Salvador Luria, Dulbecco himself, and Rita Levi-Montalcini) and had learnt some tissue culture, then in its extreme infancy. In his dingy subbasement at Caltech, he managed, after much trial and error, to hit upon a way of growing a monolayer of chick embryo cells that he could infect with virus and then stain in order to see the viral plaques. When Dulbecco achieved this tour de force for the first time and showed Delbrück, the latter was so impressed that he told Dulbecco to take particular note of the day and date; unfortunately for science historians, neither of them did. What is very clear, however, is that from the date of publication of Dulbecco's 1952 paper, "Production of Plaques in Monolayer Tissue Cultures by Single Particles of an Animal Virus," animal virus research entered the age of molecular biology, and simultaneously, molecular biology finally had a foothold in a eukaryotic system.

Cancer biology was soon to get in on the act too. Shortly after Dulbecco's development of the animal virus plaque assay, he acquired a new postdoc, a former vet named Harry Rubin. Rubin wanted to extend Dulbecco's plaque assay to viruses that did not kill their host cells, but instead caused the cells to overgrow and develop into tumours. Using the same tissue culture techniques as Dulbecco, Rubin, subsequently joined by Howard Temin, discovered that instead of plaques, tumour viruses caused the normal tissue culture cells to "transform." Virally infected cells could be seen as little foci of odd-looking rounded cells, all heaped up on each other, growing out of the otherwise flat cell monolayers. Such transformed cells could grow in culture for far longer than normal cells, and when injected into animals, caused tumours.

Temin and Rubin's focus-forming assay, published in 1958, was a gift to the cancer research community, who were in dire need of the fresh approach offered by tumour viruses. There was a growing acceptance that current research methods in cancer were not likely to yield any significant molecular information. Jim Watson, in his influential textbook *Molecular Biology of the Gene*, laid out the problem very clearly. The issue facing the biochemists and geneticists studying cancer was that because nobody was anywhere near understanding how a healthy animal cell worked, how on earth were diseased cells to be understood? In the case of cancer, a disease of uncontrolled cell growth, trying to understand how the trillions of cells in a normal body do *not* divide, and how that control can be overcome, was not even approachable using current methodology. Of the many thousands of genes in the human genome, the functions of only a few were known, so the chances of stumbling upon a crucial control gene were very small.

The solution, just as it had been for Delbrück 30 years earlier, was the reductionist approach offered by viruses. Watson pointed out that just as phages were so simply constructed that they had been used as efficient probes of basic prokaryotic biology, tumour viruses held the same promise for cancer: Their genomes were tiny but were nevertheless so powerful that an entire eukaryotic cell could be effortlessly bent to their will. To have such an effect, the tumour viruses had to be attacking the central command system of the cells, and if an intrepid explorer could follow in their tracks, they would be led to the same destination. In other words, finding what the viral proteins did would lead to the cellular mechanisms they were subverting. It was not known whether the ways in which tumour viruses caused cancer were similar to the mode of action of other carcinogens, but they were a way in to previously unknown territory that virologists, cancer biologists, and basic molecular biologists were willing and able to colonise.

By the 1960s, animal virus work had really taken off. Dulbecco's laboratory, and an increasing band of collaborators around the world, had developed methods for large-scale tissue culture. They could infect cells in sufficient numbers to purify large amounts of virus for physical and chemical analysis and could also examine what was going on in the virally infected cells. In the cancer world, it had also been discovered that tumour viruses fell into two types—those with DNA as the infectious material, and those with RNA. Of the DNA tumour viruses, two, SV40 and polyoma, were particularly prominent.

Mouse polyoma virus was discovered in 1953 by Ludwig Gross but did not acquire its current name until 1958, when Sarah Stewart and Bernice Eddy showed that it could cause multiple different tumour types in newborn hamsters, mice, and rats. SV40, purified by Ben Sweet and Maurice Hilleman in 1960 as a contaminant of the monkey kidney cell lines used to produce polio vaccines, was also shown by Bernice Eddy to cause tumours in hamsters, and could transform human cell lines in culture, although it has never been shown to directly cause human cancer. Both viruses are a similar size and shape, and their DNAs appeared to have similar properties, so they were classified as members of the same family, the papovaviruses. Their labeling as DNA tumour viruses is a gross libel: In their native hosts in the wild, they persist as endemic, harmless infections that rarely cause tumours because their hosts' immune systems keep them well under control. Only when forced by scientists to infect the wrong species at the wrong time (such as at birth, when the immune system is poorly developed), do they reveal their more sinister sides.

Polyoma and SV40 were hugely attractive to the molecular biologists studying cancer: They were small, with double-stranded genomes of just over 5000 base pairs (bp) of DNA (by comparison, the human genome contains ~3 billion bp), and they could transform cells. The existence of polyoma- and SV40-transformed cell lines, which could be grown for long periods in tissue culture, meant that not only could the changes wrought by the viruses on the cells be studied, but also large quantities of viral protein and DNA could be made; this latter was particularly important, because bucket loads of cells were required to produce enough material for the techniques available to researchers at the time.

With the tools and equipment now to hand, the molecular study of cancer began in earnest. As the new field took off, a few places in the world became centres of expertise in tumour virus research, and by extension, eukaryotic molecular biology. By the 1970s, there were two undisputed leaders in the field: the original home of the phage group courses, the Cold Spring Harbor Laboratory, and a new European interloper, the Imperial Cancer Research Fund (ICRF) in London.

ICRF Reborn: Michael Stoker and Tumour Viruses

In 1968, Cold Spring Harbor and the ICRF, after periods in the doldrums (financial in the case of Cold Spring Harbor and scientifically at the ICRF) had each just appointed new Directors. At Cold Spring Harbor, Jim Watson had taken over from John Cairns, who had heroically rescued the laboratory from fiscal disaster and put it back on its feet again, and in London, Michael Stoker had been brought in following the retirement of Guy Marrian. Stoker and Watson were good friends—they had first met in Cambridge in the 1950s—and this friendship was of great significance for the development of both the ICRF and Cold Spring Harbor. Like Watson, Stoker had realised early on the importance of DNA tumour viruses, and both men wanted to shape their laboratories around a research programme in the molecular biology of cancer. The result was a close transatlantic scientific alliance driven by a white-hot intensity of purpose on both sides, and featuring a galaxy of present and future scientific stars making fundamentally important discoveries. In this period, Cold Spring Harbor and the ICRF, nowadays renamed the Cancer Research UK London Research Institute, became the powerhouses of science that they are today. However, although the names of Watson and Cold Spring Harbor remain inextricably linked, Stoker, perhaps because he was a fundamentally nice man with a normal-sized ego, occupies a more obscure place in the pantheon of eukaryotic molecular biology.

Michael Stoker originally trained as a medic at St Thomas's Hospital in London. After war service in the Royal Army Medical Corps, he became a research Fellow of Clare College, Cambridge in 1948, and remained there for the next 10 years. His early research interests, on Q fever, a bacterial infection with flu-like symptoms, morphed in the 1950s into some of the earliest work, contemporary with Temin and Rubin's, on growing animal cells in tissue culture and virally infecting them. His research went well, and by 1957, when Renato Dulbecco came through Cambridge on a visit, Stoker was one of the people he was very keen to see. Working upstairs in a sort of hut on the roof of the Pathology Department building (for some reason, much of the really exciting work performed at Cambridge in the 1950s appears to have emanated from huts, aerial or otherwise), Stoker was very much a part of the Cambridge molecular biology scene until 1958, when he moved to Glasgow, to take up the first Chair in Virology ever created in the United Kingdom, and to direct a new virology institute. In Glasgow, Stoker further expanded his studies on cell growth in culture and moved into working with tumour viruses, doing some of the earliest research into transformation by polyoma. Today his contributions to the field have unaccountably been almost forgotten, but they were substantial and important, leading to his election as a Fellow of the Royal Society in 1968.

Michael Stoker.
(Photograph courtesy of the CRUK London Research Institute Archives.)

The brand-new, well-funded Glasgow Institute of Virology became a 1960s magnet for talented, ambitious scientists, so much so that at one point, the majority of the United Kingdom's best molecular biologists had trained in the rain and gloom of Glasgow. The influx of talent was in large part due to Stoker's qualities as its Director. Modest, unassuming, courteous as only a well-brought up Englishman can be, Stoker was an astute star spotter, and a brilliant manager of the often fractious stable of scientific thoroughbreds he accumulated. One of these, Mike Fried, summarises Stoker's talents thus: "He was so likeable that nobody bore a grudge. He could get on with everybody and get them to do what he wanted, so he was a great organiser. Nobody spoke badly of him and he was always able to work things out." In other words, under Stoker, if you had the potential to do great science, you were given the best possible chance of succeeding, however much your difficult temperament or tendencies to misanthropy tried to intervene. He was therefore an inspired choice when the Trustees of the ICRF recruited him as the new Director of the ICRF in 1968; the laboratory was in deep trouble, and it would take a scientist and administrator of Stoker's calibre to sort it out.

Stoker knew that his task in reforming the ICRF was going to be difficult. Upon accepting the Directorship, he had written to Jim Watson that "My job will be rather formidable, but I should like to see all this cancer money well spent" (letter dated 4 May 1967). The problem was that the ICRF that Stoker inherited had gone from being a central player in cancer research after its foundation in the early part of the 20th century to a rather peripheral concern, riven by arguments between the Director, Guy Marrian, and its trustees. Marrian, to his great credit, had overseen the building of new laboratories in Lincoln's Inn Fields and was in the midst of negotiating their further expansion, but his research priorities were considered outmoded, with their emphasis on traditional biochemistry, endocrinology, and chemical carcinogenesis, and the trustees were consequently unhappy with him. His health was also a cause for concern, because the stresses of the job had precipitated a heart condition. His retirement in 1968, at the age of 64, was a matter of some relief to all.

Stoker had a simple brief; to turn the ICRF into a modern research operation, and this he did, with spectacular results. However, he had some help; a good leader knows that success lies in assembling a team of able lieutenants, and Stoker had a corps of close colleagues, some from Glasgow, others hired from the world's top molecular biology labs, to form the nucleus of the reborn ICRF.

Two of Stoker's valued team would play especially important roles in the remoulding of the ICRF. The first of these was the Laboratory Manager, Bill House, an archetypically canny Glaswegian brought down from Scotland by Stoker to turn the ICRF into a place where it was as easy as possible to do world-class science. Adjutant, administrator, fixer, and general trouble-shooter, Bill completely reorganised the Central Services Department, turning it into a well-oiled machine providing reagents, solutions, tissue culture media, and sundry other services to the ever-increasing numbers of labs. He was a great believer in problem solving by example; after being told the washing up ladies were on the verge of quitting because they couldn't cope with the new regime, for three months at the beginning of his tenure he spent every morning collecting and washing glassware with them, until the system ran smoothly. Bill was the person you had to convince if you wanted new equipment or complicated media, although his efficient system of secretarial barricades meant that most often, ambushing him in the corridor was the best way of ensuring success.

Stoker had a scientific lieutenant too: Lionel Crawford, then in his late 30s, was a meticulous and green-fingered experimentalist, who, as much as Stoker, helped to forge the links with Cold Spring Harbor that were so useful to both places. Crawford's importance is evident in a letter from Stoker to Jim Watson, written in summer 1968: "I am very glad Lionel is coming to you for a few months, and hope that, perhaps through him, we can keep in touch and even set up some joint or complementary programmes. Don't keep him too long however. He is badly needed in London, particularly at the beginning, to set the proper standards. He is going to be head of a new division but will also be chairman of the group of new divisions" (letter dated 10 July 1968).

Bill House.
(Photograph courtesy of the CRUK London Research Institute Archives.)

The necessity of setting the proper standards was clear from Crawford's very first view of his new London lab space; somebody had managed to set fire to it, and it was a charred wreck. It transpired that some time previously, a technician euthanising animals with ether had finished work in a bit of a hurry, and instead of disposing of the corpses properly, had shoved them into the laboratory fridge to sort out the next day. Unfortunately, the fridge was so old and decrepit that when the compressor kicked in, it sparked and ignited the ether vapour drifting from the animals. In the subsequent explosion, the door of the fridge blew off, all the now burning flammables stored in the fridge fell on the floor, and the lab had been comprehensively trashed.

The existing denizens of the Lincoln's Inn Fields laboratories may have come to view the exploding fridge incident as a worrying omen of Michael Stoker's plans for the future, because it rapidly became clear that research in an analogous state to the fridge was destined for the rubbish heap too. In 1968, shortly after Stoker's arrival, Crawford's Cell and Molecular Biology Division, comprising five laboratories, had 17 people working in it, whereas the "old" ICRF, comprising the Endocrinology and General Studies groups, had 57 scientists and technical staff. In 1972, with the ICRF building now doubled in size after the completion of the last tranche of laboratory space, Cell and Molecular Biology contained

68 staff, and Endocrinology and General Studies had shrunk to 47. The ICRF had been well and truly propelled into the new age and was starting to accumulate the group of scientists that would make its name as a major centre of molecular biology.

In addition to the DNA tumour virologists working on polyoma and SV40, who are featured in the next chapter, Stoker invested heavily in RNA tumour virus research, a field that exploded once Howard Temin and David Baltimore independently discovered the enzyme reverse transcriptase in 1971, thereby solving the mystery of how RNA tumour viruses managed to turn their RNA genomes back into DNA so they could replicate. At the ICRF, the RNA tumour virologists, under Robin Weiss, John Wyke, and Steve Martin, worked extensively on the mechanisms by which RNA viruses were able to infect cells, but their major contributions were probably in the study of how retroviruses, a particular subtype of RNA viruses, cause cancer.

Perhaps Stoker's most spectacular hiring coup was to persuade Renato Dulbecco to come to the ICRF as its Deputy Director in 1971. Dulbecco and his wife Maureen had recently had a daughter, Fiona, and both were dubious regarding the U.S. education system, to say nothing of the political climate; the Vietnam War was still raging, Richard Nixon was President, and for left-leaning Europeans, it was all getting a bit much. Dulbecco moved to Chislehurst and stayed at the ICRF for six years, during which time he won the 1975 Nobel Prize, together with his old student Howard Temin and David Baltimore, for his fundamental work on animal viruses and tumour genetics.

Dulbecco's account of the day he found out he'd won the prize is worth relating (Dulbecco Web of Stories):

> One morning I arrived [at work], I took off my coat in my office and then I went to the laboratory. When I came back from the laboratory, I noticed that my secretary had a piece of paper in her hand. She shook it and said, "What does this mean?" I went to have a look at what she had and it was a telegram from Stockholm, from someone that I knew well, which said, "Congratulations, we'll see you in Stockholm in December," but it was mysterious, there was nothing specific—he couldn't say, as the official announcement would take place a few hours later. So, then, she said to me, "What does it mean?" so I said, "The only thing I think it can mean is the Nobel Prize." The poor woman stood as if struck by lightning. And so I said, "Before anything, I must call Maureen." I said to her, "Something has happened," and she said "Stay calm, don't say anything until we know some more." And she told me that in the meantime, she had organised a small lunch with

Renato Dulbecco.
(Photograph courtesy of the CRUK London Research Institute Archives.)

some of her friends, they were women whose children were in the same class as Fiona. And then I spoke with my friend who was the Director of the Institute and he also said, "Well, let's wait, let's see how things go." And two or three hours passed by, the time came to have lunch, I went down to the cafeteria, sat there with my friend and started to eat. At a certain point, from the door, I saw, I must say, the enormous stomach of my secretary—she was pregnant and close to the birth. And I said, "It must be something serious for you to have come down here!" And she said "There's a journalist from Stockholm who wants to talk to you on the phone." So it was true, I said this to my friend, and then things progressed from there, the telegram arrived, then the celebrations, all there in the laboratory. I phoned Maureen to tell her, and when I got home, she told me that when she received this call, she returned to her friends and her friends said to her "What's wrong, you seem worried. What's happened? Has something happened?" She said, "Yes, my husband has won the Nobel Prize!" And the reaction to the news was interesting; [her friends thought it was wonderful], but because I was living in Chislehurst, which was a somewhat privileged area, I remember speaking to people about it and they would say to me, "How is it possible for someone from here to win the Nobel Prize?" Because they thought that they were all rich and rich people don't work. How strange.

Not surprisingly, Chislehurst and the daily commute from Kent eventually proved too much for Dulbecco, who went back to San Diego to be Director of the Salk Institute in 1977.

After Michael Stoker, the ICRF went through two further Directors, Walter Bodmer and Paul Nurse, until, at the venerable age of 100, the charity that funded the laboratories was dissolved and merged with the other major cancer charity in Britain, the Cancer Research Campaign, to become a new entity, Cancer Research UK, in 2002. The ICRF laboratories, with Richard Treisman as Director, entered their final incarnation as the Cancer Research UK London Research Institute (CRUK LRI). Over the years, the changes of laboratory director have inevitably led to

The Cancer Research UK London Research Institute in Lincoln's Inn Fields.
(Photograph courtesy of David Bacon.)

changes in focus and management style, but the essential character of the place is unchanged. The laboratories are still full of creative, smart people living on the edge of the known scientific world, and the spirit of adventure that motivated their predecessors is alive and well.

Web Resources

http://library.cshl.edu/archives/genentech.html The online archives of the Cold Spring Harbor Laboratory are not only informative but entertaining.

http://oralhistories.library.caltech.edu/view/person-az/index.C-G.html Delbrück and Dulbecco's reminiscences—well worth a look.

www.london-research-institute.org.uk/about-lri/history/milestones A larger selection of scientific milestones at the ICRF and the LRI.

www.webofstories.com/play/14689?o=S&srId=237034 Dulbecco's account of his Nobel Prize day comes from the Web of Stories, an online collection of reminiscences by scientists and others, which is a terrible trap for the curious—one could easily spend a whole day there in the company of its erudite, funny, and moving subjects.

Further Reading

Austoker J. 1988. *A history of the Imperial Cancer Research Fund, 1902–1986.* Oxford University Press, Oxford. A good history of the foundation and early days of the ICRF.

Cairns J, Stent GS, Watson JD, eds. 2007. *Phage and the origins of molecular biology, the Centennial edition.* Cold Spring Harbor Laboratory Press, New York.

The classic book on the history of molecular biology, written by the people who did the work.

Quotation Sources

Schrödinger E. 1944. *What is life? The physical aspect of the living cell.* Cambridge University Press, Cambridge, UK.

Jacob F. 1970. *La Logique du vivant.* Gallimard, Paris (translated by Spillman BE. 1973. *The logic of living systems: A history of heredity.* MW Books).

Perutz MF. 1987. Physics and the riddle of life. *Nature* **326:** 555–558.

http://oralhistories.library.caltech.edu/ Interview with Delbrüuck. © 1979 California Institute of Technology.

Stent GS. 1968. That was the molecular biology that was. *Science* **160:** 390–395.

Letter dated 4 May 1967, from M.J. Stoker to J.D. Watson. Cold Spring Harbor Laboratory Archives.

Letter dated 10 July 1968, from M.J. Stoker to J.D. Watson. Cold Spring Harbor Laboratory Archives.

Dulbecco, Renato. Web of Stories; www.webofstories.com/play/14689?o=MS.

CHAPTER 2

DNA Tumour Viruses and the Fabulous Fifth Floor

In the spring of 1973, the inhabitants of the Imperial Cancer Research Fund Lincoln's Inn Fields laboratories heaved a universal sigh of relief, probably the first time they had been able to breathe out collectively in a while. For the last few years, they had been squashed like scientific sardines into half of a building, and to make matters worse, had had to endure 40 months of heavy construction work going on next door as the extension of their existing laboratory space took shape. Now, however, all was finished, and the new had been seamlessly merged with the old, but for a slight bump in the floor at the join itself. Although the assertion of the architects, Messrs Cusdin, Burden and Howett, that the brick and granite façade provided "a quiet and dignified completion to the South West corner of the square" was definitely wishful thinking (the laboratory stood out from its elegant neighbours like a sore thumb), the interiors were actually rather well designed. The laboratories were and are large and well lit, with moveable internal partition walls to allow rapid remodeling of internal space, and the wood-paneled public areas give an impression of solidity and competence, without being too showy.

The front of the Fifth Floor of the new building, with a view of the trees and flowerbeds of Lincoln's Inn Fields, was prime territory, and as such, was the domain of Lionel Crawford, the newly appointed Chair of the Division of Cell and Molecular Biology, and Director Michael Stoker's right-hand man. Lionel's molecular biology lineage was distinguished, beginning with a Cambridge PhD on DNA synthesis in phage-infected bacteria in the 1950s. During his time there, he encountered Watson, Crick, Sydney Brenner, and many other fomenters of the molecular biology revolution; crucially, Lionel also met Michael Stoker, who widened his interest in bacterial viruses to include animal ones. Part of Lionel's fascination was with the biological paradox of how viruses, such tiny and simple organisms, could wreak such tremendous havoc in the population; in the 1918 influenza pandemic, for example, it is estimated that ~5% of the entire world's population died, and a further 25% got very sick. Understanding how they worked seemed an intriguing puzzle, and when Stoker suggested that Lionel and his wife Elizabeth, an electron microscopist, move to Glasgow with him, they were happy to accept, especially because Stoker had an appetising prior assignment; he needed them to go to the West Coast for a couple of years to pick up some expertise in growing plant and animal viruses.

Lionel applied for and got a Rockefeller Foundation Travel Fellowship, despite shocking the Fellowship administrators by informing them that Elizabeth, rather than occupying the traditional wife's role of looking decorative and flying the flag for Britain, had a Green Card and intended to work too. The Fellowship, although highly prestigious, was

Lionel and Elizabeth Crawford in Cambridge, 1956.
(Photograph courtesy of Lionel Crawford.)

a pittance, but fortunately, poverty was no obstacle to having fun, and the Crawfords had a wonderful time, first in Berkeley and then at Caltech. At Berkeley, Lionel worked on tobacco mosaic virus for a while, and then migrated into Harry Rubin's lab. Rubin had moved from Renato Dulbecco's lab to Berkeley in 1958, and continued to work on Rous sarcoma virus, the RNA tumour virus he'd originally used for his focus-forming assay.

After a year in Berkeley, the Crawfords upped sticks and moved down to Pasadena, to Caltech and Renato Dulbecco. After their initial meeting in Cambridge, Dulbecco and Michael Stoker had greatly taken to each other, and Dulbecco, still confined to the subbasement, was more than happy to do his friend a favour by teaching Lionel and Elizabeth the intricacies of animal virus work. Caltech at that time was still a major site of pilgrimage for molecular biologists, so the Crawfords met Boris Ephrussi, the legendarily bad-tempered Russian who had had a hand in virtually every major molecular biology advance before the discovery of the structure of DNA. Ephrussi was determined to maintain some semblance of European civilisation in the wastes of Pasadena; he insisted on afternoon tea accompanied by civilised discourse, although after a while, tea morphed into visits to the local ice cream parlour, and the discourse frequently descended into energetically loud arguments with Howard Temin.

Working at Caltech also meant an opportunity to witness the Max Delbrück phenomenon at first hand. After an initial disappointing start, when Delbrück discovered that neither Lionel nor Elizabeth played an instrument and were not going to be able to join in his musical soirees, they became regulars on his famous camping weekends in the desert, together with an ill-assorted straggle of other international fellows. Delbrück's camping technique fell at the lower end of the comfort and amenities spectrum; all were expected to sleep on the ground, although Manny, Max's wife, did permit those who were pregnant, over 60, or both, to use air mattresses. Nevertheless, after the damp grayness of Cambridge, the desert was magical, and Lionel's love affair with it continues to this day.

When the Caltech year came to an end, the Crawfords returned to Glasgow, and Lionel set about disseminating his newly acquired skills to the cream of the United Kingdom's young molecular virologists. His own research, characterising the properties of the DNA of animal viruses, went very well, and it was during his seven years in Glasgow that he got scientifically involved with Jim Watson. Initial polite exchanges regarding reagents and experiments soon changed into friendship, and the

In the desert, 1960s. (*Left* to *right*) Renato Dulbecco, Howard Temin, Guy Echalier, Marguerite Vogt. *(Photograph courtesy of Lionel Crawford.)*

Crawfords took to spending as many of their summers as they could in the lively atmosphere of Watson's lab, first at Harvard, and then at Cold Spring Harbor. His connection to Watson and annual extended visits to the States meant that Lionel was a well-known face to U.S. molecular biologists, and by the time of Michael Stoker's appointment as ICRF director and Lionel's resulting Glasgow-to-London move, he had a reputation as an excellent scientist with an eye for the unexpected result.

Lionel's new Fifth Floor labs were a far cry from the charred fridge and burnt benches of his first visit to Lincoln's Inn Fields. All was new and state of the art, with dark-stained wooden benches, matching wood-framed shelving, and a plentiful supply of drawers and glass-fronted cupboards. The labs were furnished with a generous helping of centrifuges, water baths, shakers, incubators, and the general paraphernalia of lab life, with the biggest pieces stashed away in purpose-built equipment rooms. There was a dark room, a coffee room, office space, hoods for tissue culture and dealing with hazardous materials, and a corridor shower for those unfortunate enough to spill the hazardous materials on themselves. There was a cold room, a dark room, and a P3 lab, where especially dangerous recombinant DNA could be handled in conditions of negative pressure by suitably besuited researchers. And the whole lot was air-conditioned, a rare treat for the 1970s.

Coming out of the lifts at one end of the floor and turning left, one passed through a set of double doors, over the bump in the floor signaling the start of the new building, and turned right into the large, open-plan Crawford lab. In there, Lionel had amassed a remarkable set of young scientists, all attracted by Lionel's international reputation as the man to go to for polyoma virus research. Three, in particular—Bob Kamen, Alan Smith, and Beverly Griffin—would end up running their own independent labs and would be major catalysts of the research that transformed the Fifth Floor of the ICRF into one of the world's best DNA tumour virus research environments.

Lionel had first met Bob Kamen in Jim Watson's lab, where he was doing a PhD. A New Yorker, who favoured the bushy black beard and Bohemian casual look that was the closest thing to a uniform for male American molecular biologists in the 1970s. Bob had been hooked into molecular biology by a chance summer studentship with Harriet Ephrussi-Taylor, who had worked with Oswald Avery and was Boris Ephrussi's wife and scientific collaborator. Summers hanging out with the exotic Ephrussis, meeting their stream of scientific friends and looking after their cats when they went back to Europe, was more than enough to persuade Bob into the fold. For his PhD, because he didn't want to go to smoggy Caltech, still the obvious choice for molecular biology, Harriet set him up with Jim Watson at Harvard, although Watson, who was terrified of Harriet, was in consequence always a little wary of her protégé. After finishing his graduate work, Bob had bucked the Watson lab trend of going to Cambridge for postdoctoral work, and instead had spent some time in Zurich with Charlie Weissman, carrying on his thesis work on a bacteriophage called Qβ. However,

Bob Kamen, 1979.
(Photograph courtesy of Cold Spring Harbor Laboratory Archives.)

change was in the air: "I decided that animal viruses were the next thing. We all had this idea you had to work on the simplest possible model system. You could get your hands on the genes, you could ask good questions. I wanted to [do] a second postdoc on DNA tumour viruses, but I didn't want someone to boss me around—I didn't want to be in a lab where I was under someone's thumb. I thought of Lionel and knew that he would not be a dictatorial type. I wrote Lionel a letter and he wrote back saying please come."

Alan Smith, who'd arrived at much the same time as Bob, was cast in a very different mould. Very English, educated at Christ's College Cambridge, he had started off wanting to be an astronomer but switched over to molecular biology, infected by the excitement of what was going on in the MRC Laboratory of Molecular Biology, the new home of Francis Crick, Sydney Brenner, Max Perutz, and numerous other luminaries, and at that time the best place in the world for the structural approach to working out the nuts and bolts of life. During his PhD there with Kjeld Marcker, Alan developed a way of translating messenger RNA (mRNA) into proteins in a test tube, and using his new technique, discovered one of the fundamentals of how proteins are made; all of them begin with the amino acid methionine, coded for by the triplet DNA sequence ATG. After moving with Marcker for a short while to Copenhagen, he arrived at the ICRF, because he'd realised that it would be possible to use his very biochemical approach to study translation of the proteins being made by polyoma and other tumour viruses.

Beverly Griffin had also arrived at the ICRF from Cambridge, but her antecedents were about as different from Alan's as can be imagined. She was from the Southern United States, with a wonderful languid Southern accent, and had the sort of distracting beauty that causes mild brain disorder; susceptible scientists sometimes found themselves at a standstill in mid-sentence owing to looking at Beverly too hard. Starting off in the Chemistry Department in Cambridge, she'd eventually ended up working with Fred Sanger at the Laboratory of Molecular Biology, in the era in which he was developing the techniques in protein and DNA sequencing that would eventually

Beverly Griffin, 1979. *(Photograph courtesy of Cold Spring Harbor Laboratory Archives.)*

earn him two Nobel Prizes. Her expertise in the fundamentals of working with the DNA sequence itself was a vital addition to the diverse range of talents on the Fifth Floor.

There were two other notable inhabitants of the Fifth Floor front corridor—Guido Pontecorvo, one of the founders of the field of human genetics, semi-retired, but still an important part of the intellectual life of the ICRF, and Mike Fried. In terms of personality, Ponte and Fried were polar opposites, but together, they laid down a standard of excellence for the ICRF that kept everyone on their scientific toes. Mike, in his own words a brash New Yorker who always liked to have the last word in an argument, was an invaluable sounding board for ideas and experiments, because he was an unrelenting perfectionist when it came to science. He was another molecular biology aristocrat, having been one of Renato Dulbecco's graduate students at Caltech. Lionel, who met him during his own year with Dulbecco, remembers him as "a brilliant student—Renato had a very high opinion of him. He had a very good academic record [and] was pretty good in the lab—very careful." Mike's reputation as a major figure in the field was made during his PhD, when he worked out how to make and detect a temperature-sensitive mutant of polyoma virus; using techniques similar to those of the phage biologists working upstairs, Mike isolated the first example of a defective virus that could only transform cells at lower temperatures and showed that his mutation had to be in a polyoma protein essential for viral replication. Thanks to this pioneering work, such conditional mutants, equipped with an on/off switch operated by changing the growth conditions, became powerful tools in early studies of cell transformation by tumour viruses and also opened up tumour virus molecular genetics; using conditional mutation, individual genes could be disabled, their functions deduced, and their positions in the viral genome mapped. The relevance of Mike's work persists: conditional mutation remains an important weapon in the molecular biologist's armoury to this day.

Mike ended up at the ICRF thanks to Dulbecco's second honeymoon. After remarrying in the early 1960s, Dulbecco decided to combine a visit to his new wife Maureen's Scottish roots with a sabbatical year with Michael Stoker in Glasgow. They brought Mike with them to widen his scientific horizons, but the potent combination of his first foreign country and girls with cute Scottish accents was enough to ensure that other aspects of his life were also enhanced. He decided to do a postdoc in Britain, and then, after meeting and marrying his wife Nancy Hogg, to stay for good. His New York big city genes meant that London was always the draw for him, and thus, when Michael Stoker took over at the ICRF, Mike was very pleased to be offered a position there. By the early 1970s, he was in his own lab on the Fifth Floor, just the other side of the double doors to Lionel Crawford. As well as his own lab space, Mike also had dominion over the dark room, to get to which one had to navigate through the coffee room, juggling radioactive samples on the way in and an added cargo of freshly developed, dripping X-ray films on the way out.

Lionel Crawford, Mike Fried, Beverly Griffin, Bob Kamen, and Alan Smith: the lab heads of the Fifth Floor. The personal chemistry between some of them rivaled that found in the toxic mix of developer chemicals, dropped samples, cigarette smoke, caffeine, and stale sandwiches enjoyed by users of the coffee room. Tales of shouting matches; battles over equipment, status, and territory; the mysterious overnight repartitioning of labs; political shenanigans—all abound, but above all, the abiding memory seems to be that the Fifth Floor was a family—a large, unruly, somewhat dysfunctional family, whose members fought, made up, and fought again, but a family nevertheless, remembered with surprising fondness. And the family had one abiding obsession: to do the best science it could. Whatever the personal differences, they were put aside when it came to experiments, because science came first for everyone who worked there. As a result, extraordinary things were accomplished in a very short time.

The Fifth Floor labs were crowded places, stuffed to the brim with a polyglot collection of smart, ambitious students and postdocs keen to feed the scientific furnaces, in the hope that theirs would be the name on the next major discovery in molecular biology. More mundanely, people also came to the ICRF because they got paid well; very unusually for the time, incoming PhD students and postdocs did not have to fund themselves from external grants, because the ICRF gave them a very decent stipend. Once in, aspiring superstars found themselves in an environment superbly set up to give them every chance of success. In addition to being so well funded that virtually any experiment was possible, the ICRF, thanks to its lab manager Bill House, was one of only a small handful of lab in the world that had central facilities for media and reagents. Moreover, Lionel had set up the Fifth Floor so effectively that tissue culture cells and the viruses with which to infect them were available virtually on tap. Being able simply to go into a cold room or raid the freezer to pick up what you wanted, or to talk to Kit Osborn, Lionel's wonderful tissue culture technician, when you needed cells, gave the Fifth Floor labs an enormous advantage over their competitors, who were all toiling away making solutions when they could have been doing experiments.

Kit Osborn was a prime example of another of the ICRF's invaluable assets—its technical staff. Far in advance of the times, Bill House had had the vision to put in place a proper training programme for technicians, who could come in as school leavers and, if they had the aptitude, work their way up. Starting in the central media kitchen, technicians could earn a place in a lab, and in many cases, progress to doing a PhD and beyond, all on a paid work-study scheme. This enlightened strategy resulted in a large cohort of extremely competent people, and on the Fifth Floor at least, most labs ran at almost a 1:1 ratio of technical to scientific staff. Experiments performed by two well-trained people take an awful lot less time, and the presence of technicians who knew their way round the lab techniques and protocols ensured that even the most cack-handed student had a strong chance of succeeding.

Good staff and a great setup are solid foundations on which to build, but what made the labs buzz was the constant flow of communication. The labs were cross-fertilised by the many interactions between their occupants; new techniques, being invented almost daily at that time, spread rapidly because there was a general willingness to collaborate and cooperate. If somebody needed help with a protocol, rather than struggling along with a printed recipe, they could go and ask the person who had developed it, or at least find somebody who would know the parts of the recipe that hadn't made it into the written instructions. In this respect, molecular biology is a lot like cooking, although the results are not generally edible.

New ideas spread just as quickly. People talked science all of the time; there were coffee breaks, tea breaks, lunches, seminars, journal clubs to discuss exciting papers, and the evening ritual of going to The George, the pub round the back of the lab, which functioned as an alcoholically endowed extramural meeting room. The George was where you went for a quick pint and a sausage before going back to start your evening's work (the sausages were so horrible that one postdoc started making his own, left science, and founded a best-selling sausage company, Aidell's, in California), or where you spent long inebriated nights discussing data, theories, and gossip, or drowning your experimental sorrows in the presence of sympathetic cosufferers.

Talking is the lifeblood of science, the way work and ideas can be critically examined from all possible angles, to generate further experiments and theories or to be unceremoniously trashed. Furthermore, to find out what's going on in the outside world, there is no substitute for scientific gossip, the scuttlebutt borne back from conferences and disseminated round the lab, often weeks or months before the resulting papers appear in the journals. The annual Tumour Virus meetings, held throughout the 1970s at the ICRF's spiritual sister lab, Cold Spring Harbor, were always a major swapshop for ideas, and many collaborations were forged there at the beach, in the bar, and at all the places in between. Speculation, argument, and the occasional wild intuitive leap—these, combined with a hefty dose of rigorous logic and a profound knowledge of the field, are what drives good research, and all were present on the Fifth Floor in abundance.

Life on the Fifth Floor was made yet more exciting by the presence of migratory flocks of Americans, often from Cold Spring Harbor, come to London to check in with their colleagues, have some fun, and do the experiments that they were unable to do back home because of a cloning moratorium (in a brief fit of overzealous caution, the United States stopped work with recombinant animal DNA for 18 months in the mid-1970s). Postdocs and graduate students were not only able to hang out with the finest minds in molecular biology, but were also treated to the spectacle of some of the eminent American visitors oversampling the wares of The George, scrounging for dates to enliven their out-of-hours existence, and generally spicing the place up. The sabbatical visitors got visits from their friends too, and in some weeks, the choice of top-quality seminars was so great as to start to cut seriously into time at the bench.

Harold Varmus, future Nobel Prize winner, head of the NIH, presidential advisor, and all-round viral superstar, was one of the (more abstemious!) sabbatical visitors, arriving for a year's break from his lab in San Francisco to explore London with his wife Connie and their two small children. Harold's account of his time in Mike Fried's lab is a perfect demonstration of why he and many others enjoyed themselves so much:

> At the time of my sabbatical the ICRF was probably the best place in the world to be working on both the RNA and DNA tumor viruses of animals. Smart trainees who had remarkable careers before them—like Richard Treisman, Ed Harlow, and Adrian Hayday—were abundant. And there were many visitors—some on sabbatical, like me and Paul Neiman; some on hand to pursue romantic interests and to use recombinant DNA technology under Britain's more liberal rules, like Joe Sambrook and Mike

> Botchan from Cold Spring Harbor. The facilities were superbly set up to make efficient use of my time; the labs were close to theatre, opera, and museums; and I felt very welcome for the twice daily tea breaks, which I rarely missed. Because there was no email and the phone was expensive, I remained in only intermittent contact with my lab group, and travelled home only once during the year. As a result, I was able to do some of the best experimental work of my career and some of the best thinking regarding future work. Because we found a large old house in Cross Street, Islington, I was able to ride a bike to work nearly every day, and we were able to hire a live-in nanny who freed up more time for work and culture than might otherwise have been possible for parents of children then about 1 and 5 years old.

Here is another view, this time from an insider, Adrian Hayday. Adrian is now Professor of Immunology at King's College and a lab head at the London Research Institute, but in the 1970s he was Mike Fried's PhD student, fresh from Cambridge:

> It was flamboyant and incendiary, but I thought it was a pretty good place to train. There was a certain generation represented by people like Bob Kamen who had cut their teeth on phage, and had a certain rigour and appreciation of what very rigorous science was about. We felt similarly working on these DNA tumour viruses, because the genetics, transcription and protein analysis were so exquisitely interdigitated that you felt really good in that environment, despite the fact that some people's behaviour was not ideal!

It is hardly surprising that in this supercharged atmosphere, passions ran high. Although the hormonally driven interactions between floor members should remain in decent obscurity, the scientific arguments are exemplified by this story from one Fifth Floor inhabitant: "[My boss] and I had this big disagreement about something relatively trivial and it was *really* a big disagreement, so it got kind of loud in the corridor between the lab and his office. He marched off in a huff into the lab, and I marched off into the other door of the lab. But the lab was a rectangle, so we stormed round and kept passing each other, to the immense amusement of the other people in there. We managed three times round I think!"

People worked obsessively hard. Staying far into the night and coming in for at least some of the weekend was normal, even for the lab heads, who might be expected to want some kind of private life. The long hours were OK; the experimental work might be lengthy and sometimes tedious, but there was always somebody else around to talk to, or to nip out to The George with. In one never-forgotten incident, there was also a very hot and bothered American visitor to rescue from the high-security P3 lab in the middle of the night—how he had managed to lock himself in there in the first place was anyone's guess. And once work was finished, there were the parties, although curiously, participants seem to be a little hazy on exactly what made them quite so legendary.

In the early 1970s, knowledge regarding SV40 and polyoma was still fairly limited, and there was a lot of scope for discovery. The size and shape of the viral capsid, the protein coat that enclosed the infectious viral genome, was known, and the double-stranded DNA genome had been shown to be circular. There was some understanding of what happened during a productive viral infection, in which a virus enters a cell, replicates, and then bursts (lyses) the cell, releasing hundreds of progeny. There were two stages, early and late, classified as happening before or after the viral DNA was replicated, and different mRNAs and proteins were made at each time. Late-stage proteins were to do with making the structural bits of the virus, fitting out the newly replicated DNA genomes with capsid coats before their release by lysis. In contrast, in the early stage of infection, the viruses' main concern was to persuade their host cells to fire up DNA synthesis. This they did by trickery; because the viral genomes were limited in size by the rigid capsids in which they were contained, they had evolved to carry a cargo of genes able to activate the host's own replication machinery, which could be fooled into copying the viral DNA in addition to its own.

The early stage of infection was most interesting to those concerned with how DNA tumour viruses caused cancer, because the early genes were the only things the virus switched on during transformation, the lab-induced state roughly analogous to tumour growth in animals. Viral infections leading to transformation were always nonproductive—there were multiple cellular changes leading to uncontrolled growth, but there was no lysis and no production of more virus—indeed, the virus disappeared.

The mystery of the missing virus in tumour-virus-transformed cells and cancers was cleared up in the 1970s, when it was found that the viral genomes forced all or part of themselves into the host cell's own chromosomes, thereby guaranteeing their propagation every time the hapless cell divided. In an interesting insight into the mind-set of the tumour virologists, who tended to view everything from the standpoint of the viruses, this rather violent process had been given a positive spin and was called integration (similarly, the cellular Armageddon of lysis was described as "liberation of infectious particles").

In transformation, as early but not late events happened, an early protein was the obvious culprit for corrupting the cell. The likely villain had a name, T (for "tumour") antigen, but only a nebulous identity because of the method of detection. When its xenophobic immune system finds a developing tumour, an animal with a virally induced cancer makes antibodies directed against tumour antigens, the foreign-looking bits of the tumour cells. The tumour antigens are likely to derive from the viral proteins because these are the most alien component of the tumour and will elicit the biggest immune response. Antibodies circulate in the bloodstream, and thus extracting blood from tumour-bearing animals produces a tumour-specific antiserum, which can be used for experiments.

Quite early on, in the 1960s, it was realised that when used to label cells in tissue culture, tumour antisera went straight for the tumour cells, ignoring the normal control cells. Used at such low resolution, the antisera were like burglar alarms, able to detect the presence of T antigen on the tumours, but unable to tell what it was; T antigen could have been a posse of tiny tap-dancing penguins, for all the antisera could tell, although admittedly this was unlikely. Whatever the true nature of T antigen, its ubiquitous presence on tumours meant that it was the obvious suspect as the inducer of the cancer; how it might do this was, however, unknown. In line with the virus's need to subvert cellular DNA synthesis, there was speculation that T antigen's normal function might be to switch on the cell's replication machinery, perhaps by suppressing a vital negative regulator—a bit like taking the handbrake off in a car—and in cancer, this function was never switched off, leading to tumours.

The basics of DNA tumour virus biology were therefore established: DNA tumour viruses entered cells, most often replicating and killing their hosts, but sometimes integrating to cause tumours, probably by some mechanism involving the mysterious T antigen. But how was all this accomplished? There was almost no understanding of mechanism, and the matters for investigation for the Fifth Floor and their competitors and collaborators in the DNA tumour virus lab around the world were fairly obvious, if a bit daunting. Nobody knew how many genes the viruses had, what proteins these genes encoded, and what the proteins did. Nobody knew how the genes were organised on the viral DNA or what their DNA sequences were. Nobody knew how the viruses managed to squash all their DNA into the claustrophobic constraints of the viral capsid, and nobody knew how T antigen managed to induce cells to replicate their DNA uninhibitedly to form tumours.

With such a comprehensive to-do list, it was very good that as the 1970s progressed, the molecular biology methods repertoire underwent an exponential expansion, from a small number of low-resolution, highly radioactive, potentially toxic techniques to a much larger number of higher-resolution, highly radioactive, potentially toxic techniques (advances in safety had to wait until well into the 1980s). The turning point was Hamilton Smith's purification in 1970 of a restriction enzyme, one of a large group of bacterial enzymes capable of cutting up DNA by recognising and snipping open a specific DNA sequence. The following year, Dan Nathans and Kathleen Danna showed that SV40 viral DNA could be cleaved by Smith's enzyme preparation into defined fragments. With the purification of more and more restriction enzymes, each with a specific cutting sequence, DNA could now be chopped up into small pieces, and the pieces ordered relative to one another by doing single or double enzyme digests. (It turned out that Smith's original prep actually contained two enzymes, so Danna and Nathans's paper inadvertently showed the fragments generated after a double, not a single digest of SV40.)

With the availability of restriction enzymes, mapping viral genomes became possible, and after Herb Boyer and Stan Cohen's demonstration in 1972 that a piece of restriction enzyme–digested DNA could be introduced into a plasmid, a small circular DNA parasite able to replicate in bacteria, cloning was born. Once put back into bacterial cultures, plasmids carrying cloned inserts could replicate their DNA very efficiently, meaning that a small amount of DNA could be amplified into the large amounts needed for the experimental techniques of the day.

In tandem with these advances, methods were developed to analyse DNA, RNA, and protein more easily. The major advance was the invention of gel electrophoresis. Up to the late 1960s, DNA, RNA, and proteins could only be analysed at extremely low resolution by electron microscopy, or at slightly higher resolution in laborious sedimentation experiments in which molecules were separated on the basis of their size by spinning cell extracts in ultra-high-speed centrifuges for many hours, sometimes days. Gel electrophoresis cut separation times to 2–3 hours at most; it had far higher resolution, down to a single base in the case of DNA and RNA, and for the first time, molecules could be "seen" as stained or radioactively labeled bands, which could then be analysed further. Gels could be tailored in size, thickness, and chemical composition, and were easily made by pouring liquid gel mix into flat moulds and allowing it to set. Molecules were separated, still on the basis of size, by being pulled through an electrical field. A gel was submerged in a suitable charge-carrying solution between two electrodes, and when the current was switched on, molecules migrated from their loading slots at the negative end of the gel towards the positive electrode at the other end. How far they went depended on how big they were; large molecules moved slowly, and small ones wriggled through the gel matrix far faster. After they'd been run, gels could be stained with DNA or RNA fluorescent dyes, visible under UV light, or, if molecules had been radiolabeled, the gel could be wrapped in Cling Film and exposed to X-ray film, revealing the radioactive bands; for example, in the case of Danna and Nathans's SV40 restriction map, there is a ladder of 11 bands.

In 1974, gel electrophoresis became even more useful when Ed Southern worked out that gels could be blotted wholesale onto special membranes made from nitrocellulose, which could then be probed with radiolabeled ("hot") DNA or RNA. Southern blots revolutionised mapping; because it was now possible to pinpoint exactly which restriction fragment a probe recognised, the location of specific DNAs and RNAs within genomes, whether viral or otherwise, could now be established. Southern blotting was followed in 1977 by a method for transferring RNA rather than DNA from gels, inevitably dubbed "northern blotting," and in 1979, by "western blotting," using antibodies to probe protein blots ("eastern blotting" doesn't exist, but not from want of trying; many researchers have tried to appropriate the name for a multitude of new techniques, but none have succeeded in making it stick).

Nowadays, in a world of endless kits, standard equipment, and nonradioactive labeling methods specifically designed so that even a precociously talented toddler could do molecular biology without poisoning herself, it is hard to appreciate how difficult and potentially hazardous it was to do experiments then. Almost all of the equipment and reagents were laboriously homemade. At the ICRF, enzymes were cooked up in vats in the second basement, and media, agar plates, and some reagents were made in the kitchens. Apart from that, you were on your own and had to make up solutions and reagents yourself (one extraction reagent was abbreviated to 2FC, ostensibly an acronym for its ingredients, but mostly because it was Too F***ing Complicated to make). Gel boxes and equipment were made in the workshop to patterns specified by the scientists, and the skilled craftsmen working there were highly valued. Because everything had to be performed on a very large scale, the limits of resolution of most methods still being fairly low, labs were full of *stuff*; cupboards in the labs and corridors were jammed full of glassware; tissue culture paraphernalia littered the hallways and shelves; gel boxes and the power packs to run them clogged the benches; and measuring cylinders, pipettes, pipette tip boxes, tubes, Petri dishes, and flasks of all possible sizes were everywhere. In the tiny offices, even a slight tendency to untidiness resulted in chaos, with tottering piles of papers, journals, and, in one case, a spent matchbox collection, infringing every known Health and Safety regulation. The clutter and general dinginess were not helped by the miasma of cigarette smoke and the presence of senior technician Bob MacCormick's dead pheasants, often to be found hanging in the cold room after a weekend's successful shooting.

Richard Treisman, now Director of the London Research Institute, started as a graduate student with Bob Kamen in 1977. His recollections of a couple of the techniques he had to do during his four years at the coalface give some idea of just what was involved. Firstly, preparation of RNA:

> To make an RNA prep, you would do a viral infection which would involve infecting a hundred 9 cm dishes of cells. Then you'd have to scrape them all with rubber policemen [an exotic name for a prosaic tool for getting cells off tissue culture plates] and lyse them. Everyone was using buckets of stuff because everything was low specific activity—there weren't pinpoint methods for analysing RNA. You'd have to make polyA+ RNA [a.k.a., mRNA] and because that was only two percent of the total cellular RNA if you wanted to quantify it accurately you'd have to make a lot of it. It was a very blunt instrument.

And for making radioactive DNA probes:

> We got ^{32}P (a highly radioactive phosphate compound, confusingly referred to orally as P32) from Amersham. You'd get it on a Friday, and it would come in at lunchtime, so you could dump it on to your cells after lunch. There were just tons of hot phosphate! To make any probe, we

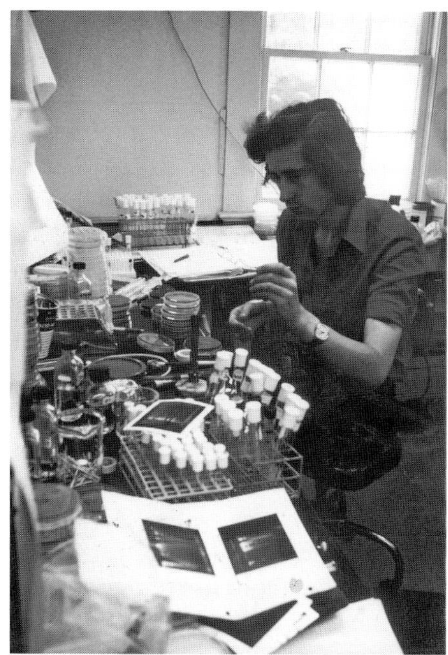

Richard Treisman, ca. 1975.
(Photograph courtesy of author.)

> would radiolabel DNA in vivo. For a DNA prep, you'd take four dishes of cells, infect them with polyoma, and then you'd put 10 milliCuries of ^{32}P on a dish. That would be your Friday afternoon. You'd leave the cells for three hours from 1 to 4pm, then you'd have to make supernatant. So you'd fart about with these red hot plates making your lysate behind screens, and dump the lysate in Corex [glass] tubes. The supernatant had to extract at 4°C for three hours, which conveniently coincided with the opening of the pub, so you'd let that cook for a bit, and after a couple of beers you'd go back and spin out the gunk and take the supernatant and run it over a ion exclusion column. You'd get the viral DNA out and you'd then dry it down and you had some hot DNA that you could use for a hybridization experiment. Labelling fragments was a big pain in the bum.

Sometimes the drying down procedure was a little difficult: "We had to label up a DNA fragment with ^{32}P and dry it down in a desiccator, and we were doing it on a vacuum water pump. While I was drying it down, there must have been a fluctuation in the water pressure so the vacuum got stronger than the pump could generate. It sucked the entire water stream back into the desiccator, which filled up to the brim all over the nicely purified DNA, which we then had to [re]purify from limescale because we were so desperate. It was a disaster but it worked very well."

And all this was repeated day after day, week after week, with astonishing persistence on the part of the investigators.

What exactly was achieved by the Fifth Floor labs? The short answer is that they helped build the foundations of modern biological science, in a series of contributions that were intricately tangled up with contemporaneous work performed in the powerhouse labs of the United States. They then went on to populate the world's labs with their trainees, many of whom have risen very high. They were, to paraphrase Michael Stoker in his valedictory Director's Report, riding a bandwagon, but playing at the front. The field moved fast, and they had helped establish it and continued to drive it.

Early on, an informal arrangement on how to carve up the work was agreed, with each person playing to his or her strengths: Lionel would continue with his work on viral DNA, its structure and how it replicated, with a bit of protein work on the side. Mike would do the molecular genetics, making mutants and using them to locate viral genes and determine which were important. Beverly (who had moved shortly after her arrival into Mike's lab space on the other side of the double doors) would map and sequence the polyoma viral genome, working out where genes were located and what they encoded. Alan would work on viral proteins, and Bob would work on viral RNA. The overarching issue of how the viruses caused cancer would be tackled by determining which bits of their genomes mattered for transformation and by studying T antigen. There was complete sharing of information. Beverly and Mike's genetic and sequencing data were fed hot off the presses into Bob, Alan, and Lionel's work on RNA and protein, and vice versa. And if, as was sometimes the case, the lab heads were currently fighting about something, the information flow still continued through their postdocs and PhD students. The result was phenomenally rapid progress on all fronts.

Mike and Beverly's lab, working first together and then rather more separately as their interests diverged, made the first restriction map of polyoma DNA, following closely on the first SV40 map of Danna and Nathans. Mike expanded his collection of mutant viruses to those having more subtle defects, and by using a neat technique to chop and change bits of DNA, he was able to use the polyoma restriction map to pinpoint where the defects lay in the virus and therefore where important genes were located. Until cloning and sequencing became commonplace, the polyoma mutants he generated were also the biological backstop for everyone else on the floor; predictions of which proteins were implicated in transformation could always be checked, because the protein concerned had to be mutated or absent in the corresponding viral mutant. In parallel, Beverly ramped up her DNA sequencing, and in a series of high-profile papers in *Cell* and *Nature*, published the complete DNA sequence of polyoma virus at the end of the 1970s. Polyoma, and SV40, completed by Sherman Weissman's lab at Yale in 1978, were the largest pieces of continuous DNA sequenced at that time, an enormous technical achievement. The number of bases that could be sequenced in one experiment numbered in the tens, and to ensure accuracy, a sixfold redundant coverage of the polyoma genome meant that the total number of bases sequenced was more than 30,000. The sequence had to be read from X-ray films (naturally, the experiments relied on lots of screaming hot ^{32}P to label the DNA) and pieced together entirely by eye, a jigsaw puzzle of epic proportions. The whole process was a mix of long hours of tedious repetitive work combined with the need for extreme accuracy to ensure that every base of the sequence was correct.

DNA tumour virus investigators were handed the blueprints of their viruses with the polyoma and SV40 sequences, twin Rosetta stones allowing the mysteries of the viral genomes to be resolved. Regions occupied by genes could be identified, and the sequences of the proteins encoded by the genes could be deduced, by translating the triplet codons specified by the DNA into the run of amino acids making up the protein. The sequences of the mutant genes mapped in Mike's genetic screens could be compared with their normal counterparts, and the amino acid changes causing the mutation identified; sequence could be correlated with functionality, and guesses made regarding what the proteins did, and what was going wrong.

The two virus sequences were also the start of something much bigger: they were the first demonstration that it was possible to assemble a complete genome sequence and that the information gained from the sequence was valuable. This led to the sequencing of much larger viruses, which, in turn, produced the exponential explosion in DNA sequencing technology and analysis culminating in the unveiling of the draft sequence of the human genome in 2001. Today, sequencing DNA is a triviality, performed by robots that can easily sequence millions of bases in a single day.

In an extension of his transformation mapping, Mike also started on some of the earliest experiments cloning bits of polyoma virus, inspired by his collection of sabbatical visitors, who brought with them Boyer and Cohen's DNA cloning protocols. Mike, being a very visible star in the polyoma firmament, was a prime destination, and in addition to Harold Varmus, hosted, amongst others, Phil Sharp, Joe Sambrook, and Mike Botchan. Joe Sambrook came even more frequently after he fell in love with Mary Jane Gething, who was working downstairs in Mike Waterfield's lab.

Mike's mastery of cloning resulted in a very significant *Nature* paper in 1979, which resolved a key safety issue in the cloning of eukaryotic DNA. As mentioned above, several of the ICRF's transient American scientists were there in part to get around the cloning ban that the U.S. biological science community had voluntarily put in place in the mid-1970s. Cohen and Boyer's demonstration of cloning, with its potential to move genes from one organism into another, had provoked extreme alarm in some of their fellow scientists, with inevitable fears of spreading dangerous genes through the human and animal populations. Even when cloning resumed, the suggested restraints mean that any cloning experiment involving eukaryotic genes had to be performed under extraordinarily stringent containment, which, being expensive, meant many experiments could not be performed at all. Although this, as pointed out by John Cairns, by now head of the ICRF's North London outstation at Mill Hill, meant that the less competent had a perfect excuse for not doing hard experiments, it was a real nuisance, especially as it became increasingly clear that the regulation was excessive. Mike and his collaborators put the lid on the controversy by cloning the complete polyoma genome into a bacterial plasmid, introducing it into mouse cells, and showing that in this form it was not infectious and could not spread.

Joe Sambrook and Mike Fried, 1974. *(Photograph courtesy of Cold Spring Harbor Laboratory Archives.)*

The problem of what mRNAs and proteins were made by SV40 and polyoma was occupying Bob, Lionel, and Alan, on the other side of the double doors. Their large, shared lab had an interesting appearance. At the Smith end, it was cluttered, but nevertheless forensically tidy, reflecting the extreme neatness of Alan and his head technician, Rob Harvey. At the Kamen end, it was a chaotic mess, where new experiments were begun not on a cleaned and empty bench, but in a small space made by pushing to one side the heaped detritus of the previous day's work. In the centre, Lionel's bay provided a convenient buffer zone, valiantly fighting off the encroaching tide of junk. Lionel, very much in the Cambridge scientific tradition, did his own experiments and facilitated those of his lab, only putting his name on papers if he had contributed directly to the data. His work on DNA structure and replication continued to attract attention, notably when he located the polyoma origin of replication in 1973, and broadened to include more biochemical analyses of the proteins involved. With a young Bruce Ponder (just retired as Director of the CR-UK Cambridge Research Institute, but then a newly qualified MD), Lionel produced a trilogy of papers on how viral DNA was wrapped for packaging in the capsid by host proteins called histones, and how this differed from normal histone packaging of host DNA. On the transformation side of things, to try to develop a better way of turning viral mRNA into functional protein, Lionel also hired Ron Laskey, folksinger extraordinaire and a future giant of the DNA replication field. Laskey, who had been a student in John Gurdon's lab, spent two years with Lionel looking at whether the eggs of the clawed frog, *Xenopus laevis*, were a suitable venue for translating viral mRNA into functional viral protein. The proteins the eggs made were full length and functional, an important advance, but the dream of trying to get transformation in frog eggs by injecting different mRNAs was, sadly, never realised.

On either side of Lionel, despite their differences in lab hygiene, Bob and Alan got along well and collaborated extensively, Bob contributing his expertise in making RNA, which Alan could then plug into his cell-free systems and translate into protein. Bob's literary style also benefited: "[Alan] was much more organised than I was—he could actually write papers. I couldn't do first drafts and he did first drafts very easily. Alan would look at all my adjectives and take them down one level of superlative!"

Cell-free translation, once methods of making viral mRNA able to programme the system to produce protein were worked out, yielded a rich vein of experimental data. The Smith and Kamen labs were able to figure out where in the polyoma genome the viral mRNAs came from, and Bob had his first lesson in paper writing from Alan when they jointly published a *Cell* paper showing the genomic location of the viral capsid proteins.

Alan Smith, 2002.
(Photograph courtesy of CRUK London Research Institute Archives.)

The Smith lab, with its connections to Cambridge and the Laboratory of Molecular Biology, attracted some high-calibre trainees, amongst them a young Tony Pawson, whose PhD project there was to look at cell-free translation in Rous sarcoma virus. Tony, erstwhile Director of the Sam Lunenfeld lab in Toronto, Companion of Honour, and winner of multiple prizes and medals, including the Kyoto Prize, the "Japanese Nobel," thought the ICRF was "... a daunting challenge for a young 21-year old. On the other hand, it was the most stimulating and thought-provoking environment that one could possibly imagine. The cloning of recombinant DNA was just starting, as was the tentative identification of genes and proteins that might be causally involved in the development of cancers. I shared a bench with Ed Ziff [one of the pioneers of the transcription field] who had just developed an early approach to DNA sequencing, and he was unfailingly generous with his time and advice."

Tony, like Harold Varmus, was also a teabreak enthusiast: "Best of all, in the fine English tradition, everything stopped for tea in the morning and coffee in the afternoon, during which all the latest ideas were hotly debated and my views were solicited as though I should know what I was talking about" (Pawson 2008).

Down at the transcription end of the lab, Bob soon realised that working on the mRNA of tumour viruses, rather than being a brief prelude to understanding virally induced tumours, was a full-time undertaking, full of technical and conceptual pitfalls. It all started off well; using low-resolution techniques, made even more difficult by the tiny amounts of RNA he was trying to detect, he was the first to locate the early and late mRNAs of polyoma on the viral genome, and to work out that one of the two complementary strands of DNA in the viral double-stranded genome was used to transcribe early mRNA, and the other strand made the late mRNAs. However, things went pear-shaped when he tried to map polyoma late mRNAs more precisely, by doing Southern blots using radioactively labeled viral mRNA to probe the ladder of restriction fragments produced by digesting the polyoma genome. When Bob chopped polyoma DNA up and did this experiment using late polyoma mRNA as the labeled probe, bands lit up on the film, sure enough, but they looked all wrong. The RNA bound to restriction fragments that Bob knew from Mike and Beverly's polyoma genome map were not next to each other, which was totally weird. In bacteria and phages, where all the transcription work had been performed to date, mRNA was transcribed from one long unbroken strand of DNA, so one could match it exactly to the region of DNA from which it came. Therefore, in Bob's Southern blots, the polyoma mRNA should only hybridise to adjacent DNA fragments, not jump around like a demented salmon on a fish ladder. Bob abandoned his half-written paper to an upper shelf and got on with other things for a while.

The puzzle was finally solved by the ICRF–Cold Spring Harbor connection, in the person of Joe Sambrook, who showed up in Bob's office one day in 1977 with some advance notice of a major scientific breakthrough by Phil Sharp at MIT and Rich Roberts at Cold Spring Harbor—the discovery of splicing. Both Sharp and Roberts were working on adenovirus, a larger DNA tumour virus that can cause human cancer, and both had now shown that adenovirus late mRNA molecules were made in an entirely novel way. As in bacteria, the RNA was transcribed as one long molecule, but then there was a new, eukaryotics-only twist to the story: the long RNA was then spliced. Exactly as one edits out an unwanted scene in an old-fashioned film reel, the cell chopped loops out of the precursor RNA to produce the final, smaller molecule. Therefore, as Bob had been seeing back in London, mapping of the final product would make it look as though the RNA was being made from bits of unconnected DNA.

Bob was initially reluctant to believe what he was hearing that day: "Sambrook was a bit of a character, funny but with an edge—he looked a bit like Mick Jagger. He started telling me about this splicing story and I thought it was an elaborate hoax to get me to bite so he could call me a fool. So I called him on it, and he exploded and said 'you idiot!' So not only did I miss it but I didn't understand it!" Bob was not alone in his surprise that splicing existed, or in his chagrin at missing it; there were lots of unexplained transcription results lying on the shelves of multiple labs at the time. Sharp and Roberts won the 1993 Nobel Prize in Physiology or Medicine for their insight, leaving their colleagues in the field to contemplate what might have been had they looked at their strange results in a different light.

After splicing was discovered, eukaryotic mRNA studies were a lot easier to decipher, but also became a lot less simple. Once it was proven that splicing was not just a viral peculiarity, but happened in all eukaryotes, it was clear that transcription of a eukaryotic genome was incredibly complex, with multiple mRNAs being made from the same stretch of DNA, mRNAs starting and stopping in different places, and differential splicing galore. The study of the DNA sequences controlling transcription and the proteins that bound to those sequences to direct the process, became and remains a field in its own right, and transcription research continues to produce unexpected explanations for mystifying data.

For SV40 and polyoma, splicing was the key to understanding why the viruses were so powerful. In getting around the constraints of being packed into a small capsid, both viruses had come up with an ingenious way of maximising their impact on a cell. Like tiny tardises, their genomes had the potential to encode far more than had been thought possible, because multiple differentially spliced mRNAs could be transcribed from the viral DNA.

One of the DNA tumour virus puzzles solved by the discovery of splicing concerned the nature of the viral T antigens. As discussed above, the T antigens of polyoma and SV40 started off as abstract definitions for the way tumour-specific antisera could light up virally induced tumour cells in immunocytology assays. People assumed that T antigens existed and inferred their function in cancer from the evidence (transformation) left at the scene of the crime, but it was impossible to tell whether T antigen was an oncogenic gang or a lone perpetrator. This all changed when, using the tumour-specific antisera as bait, people started to fish out T antigen from radioactively labeled cellular protein soups. This technique, immunoprecipitation, when combined with gel electrophoresis, meant that the T antigen "catch" could be seen by running the immunoprecipitated proteins on a gel. For both polyoma and SV40, there was always a protein band hovering at around 90–100 kDa (kilodaltons) on a gel (molecular biology operates on a tiny scale, and needs its own tiny units;

a kilodalton, abbreviated kDa, has a mass of ~1.66×10^{-24} kg). Because this was about the maximum size that the whole early region should be able to encode, the 100-kDa protein reigned supreme as T antigen for several years.

The belief that T antigen was just the one 100-kDa protein was abandoned in 1977, in the same year as splicing was discovered, when both SV40 and polyoma were shown to have two T antigens. The original 100-kDa protein was renamed large T, with its new smaller companion unsurprisingly christened small T. SV40 and polyoma small Ts were both ICRF finds, one discovered by Lionel, off on sabbatical on the West Coast in future ICRF group leader George Stark's lab, and the other by Yoshi Ito, a postdoc in Renato Dulbecco's fourth-floor lab. (Ito was to become another star of the cancer field and was also a major force behind Singapore's transformation into a world-class centre for biomedical research.) In the case of SV40, Lionel and his Stanford colleagues also scored a first regarding small T's origins, when they proposed that it shared the same front end ("amino terminus") as SV40 large T but had a different back end ("carboxyl terminus"); in all but name, they had shown that splicing could occur, and their data slightly preceded Sharp's and Roberts's. The relationship between large and small T was rapidly confirmed, at the protein level by Alan Smith's lab at ICRF and Carol Prives and her colleagues at the Weizmann Institute in Israel, and at the mRNA level by Phil Sharp and Arnie Berk at the Massachusetts Institute of Technology (MIT). The two proteins are made from the same genomic sequence but are splice variants: SV40 small T mRNA is unspliced and only encodes 174 amino acids before a stop codon is reached, and translation, and hence small T protein, terminates. Large T mRNA also starts off at the same place but is spliced to remove the sequences unique to small T, including the stop codon; it therefore makes a far larger final product 708 amino acids long. The discovery was important not just to the SV40 field; Sharp and Roberts's adenovirus splices were in an untranslated part of the RNA, so the large T/small T result was the first example of splicing in a coding region. It illustrated just how exquisitely precise the process had to be; an exact base-to-base join always had to be made at the splice junction, because otherwise the proteins translated from the mRNA would be gibberish.

Much the same process was going on in polyoma, with an added complication to the story. Ito and colleagues in the Dulbecco lab showed in 1977 that polyoma not only had a large T and a small T antigen, just as for SV40, but that it also had a middle T, that, unlike large and small T, which lived in the nucleus, was found stuck in the outer plasma membrane of the cell. The following year, Ito and protein chemist John Smart, and their competitors Mary Anne Hutchinson, Tony Hunter and Walter Eckhart at the Salk Institute in San Diego, published that all three T antigens shared the same amino terminus, middle T and small T had the same middle part, but the carboxyl termini of all of the three proteins were unique. In a further tribute to viral ingenuity (and also to the technical excellence of the people involved; it was a very fiddly experiment), the Kamen lab showed in 1981 that the unique region of middle T and part of large T were actually encoded by exactly the same stretch of DNA, using two different reading frames. (Because the amino acids making up a protein are coded for by triplets of DNA sequence, there are three possible ways of decoding, "reading," an mRNA molecule to make protein, depending on whether one starts in frame 1, frame 2, or frame 3; for example, "the fat cat ate the bat," if read in frame 2 becomes "t hef atc ata tet heb at." Polyoma, rather more cleverly than me, makes sense in both frames, not just the first.)

Middle T was an important find, because it was the oncogene, the cancer-causing gene, of polyoma. Yoshi Ito's paper had hinted this was the case, because middle T was not made or was defective in mutant viruses unable to transform cells, but it took a cloning tour de force executed by Richard Treisman in the Kamen lab to establish middle T's status as chief villain conclusively. Using techniques so new that they barely existed, Richard poked a hole in the polyoma genome such that only middle T could be made, and went on to show with help from his then girlfriend, Uli Novak, a student with Beverly, and Bob's technician Jenny Favaloro, that the middle T–only virus could still transform cells. Joining in on the act, a collaboration between Beverly Griffin, Alan Smith, and Mike Fried showed that middle T was associated with a protein kinase activity, which proved to be the key to its activity. Protein kinases stick phosphate groups on amino acids, causing changes in how proteins can interact with each other, and affecting how signals are passed around in the cell. Phosphorylation, and its opposite, dephosphorylation, is one of the major regulatory mechanisms used by cells; the reversibility of the reaction makes it a versatile and rapid on/off switch, and finding that middle T was capable of overriding such a switch was tremendously exciting. The work was published in back-to-back *Cell* papers in 1979 with two other labs, Tom Benjamin's at Harvard, and the Salk grouping of Hunter, Hutchinson, and Eckhart. Tony Hunter's paper additionally showed that the protein kinase activity was specifically aimed at tyrosine residues, which had never been observed before. Alan, after leaving the ICRF in 1980 for the National Institute for Medical Research at Mill Hill, showed with Sara Courtneidge that the tyrosine kinase activity came from a cellular protein called c-Src, with which middle T associated.

To finish the middle T story, c-Src was the first of many proteins with which middle T was shown to interact, and all these multiple interactions mean that middle T sits at the centre of a web of signaling pathways and turns them all on, driving cells into uncontrolled replication and thus causing cancer. Using middle T as a probe for derangement of cellular function proved to be hugely important for cancer research, more than fulfilling the DNA tumour virologists' hunch that their viruses were a simple way into a complex system. Middle T strikes right at the heart of a cell's regulatory pathways, and all of the proteins and interacting pathways first found using middle T are universally important in controlling normal and cancerous growth.

Middle T had one more trick up its sleeve, another clue as to how oncogenes worked. Bob Kamen, once he had finished the mountainous task of mapping all of the polyoma mRNAs, turned to studying the function of polyoma proteins, in particular, the relationship between the three T antigens. He set up a collaboration with François Cuzin in Nice, looking at what each T antigen did, using his original middle T–only virus, and two others, engineered to express just small T or just large T. Middle T, sure enough, could transform cell lines by itself, but to transform primary cells taken straight out of an animal, rather than cells that could grow indefinitely in culture, it required help from large T. In contrast, large T by itself could immortalise the primary cells, turning them into cell lines, but could not take the final step of transforming the immortalised cells. This idea, that one gene was required for the first step towards cancer, and then a second was needed for the next step, gave rise to the cooperating oncogene hypothesis, which would make the names of MIT scientists Hucky Land, Luis Parada, and Bob Weinberg shortly thereafter. As Bob Kamen remembers: "I gave a talk at a meeting in Houston which was the first public disclosure of our work. Weinberg was in the front row and he came up afterwards and said 'this is amazing—it explains some work we've been doing.' Weinberg went straight back to his lab and convinced them that polyoma was doing something that mimicked cancerous changes caused by cellular mutations." In Weinberg's case, a cellular gene called c-*myc* could be forced to provide the stimulus for the cell to replicate indefinitely, and overexpressing another gene, c-*ras*, transformed the myc-immortalised cells. Land, Parada, and Weinberg's paper, "Tumorigenic Conversion of Primary Embryo Fibroblasts Requires at Least Two Cooperating Oncogenes," was published in *Nature* in 1983 and is a much-cited classic, presaging our modern understanding that there are multiple steps from normality to cancer, and different oncogenes play different parts. Bob is a little regretful that he and Cuzin didn't capitalise more on their work, which was published the year before Weinberg's, also in *Nature*: "Weinberg certainly knows how to publicise! He always acknowledged us, but he was just a better publicist."

With all the kerfuffle on the Fifth Floor surrounding polyoma, it was inevitable that the SV40 camp started to wonder whether their virus had a middle T antigen too. SV40 large T did some suspiciously oncogenic things, like stimulating cellular DNA synthesis, but perhaps there was more to it than that; perhaps it was just a decoy, or perhaps it was only part of the story. It would be great to find an SV40 middle T, but the problem was that the immunoprecipitation technique, the method of cellular fishing described above, was, to be frank, pretty rubbish. There was certainly a slew of other bands that were pulled down in immunoprecipitates at the same time as SV40 large T, but most of them were probably chopped up bits of large T or background contaminants, there because the anti-tumour antisera could sometimes be alarmingly nonspecific. Yoshi Ito had succeeded in finding polyoma middle T by spending a lot of time tinkering with his immunoprecipitation conditions, and something similar was needed for SV40. Lionel, especially, was aware of this, and started to think very hard about how to proceed, talking at length to the hard-core immunologist Av Mitchison at the ICRF Tumour Immunology Unit just down the road at University College. Av, although sympathetic, was not very helpful. Fortunately, help was at hand in the person of Av's PhD student, a promising young chap with a knack for immunochemical troubleshooting. To cut to the chase, Lionel and his new recruit would find that there was no SV40 middle T. What there was instead, for all of the marvellous doings of the rest of the Fifth Floor, would in the end eclipse them completely.

Web Resources

www.dnalc.org/resources For short, informative, watchable videos on the basics of DNA, cloning, restriction enzymes and molecular biology principles in general, see the Cold Spring Harbor DNA Learning Center website.

Further Reading

Angier N. 1988. *Natural Obsessions.* Houghton Mifflin, New York. A cracking account of life in Bob Weinberg's lab at the time of the oncogene cooperation hypothesis.

Cheng J, DeCaprio JA, Fluck MM, Schaffhausen BS. 2009. Cellular transformation by Simian Virus 40 and Murine Polyoma Virus T antigens. *Semin Cancer Biol.* **19:** 218–228.

A review of polyoma and SV40's many talents as tumour viruses

Varmus H. 2010. *The art and politics of science.* WW Norton, New York.

Harold Varmus appears only transiently in this chapter, but his autobiography is a fantastic insight into the life of one of molecular biology's most eminent and interesting personalities.

Quotation Sources

Pawson AJ. 2008. Kyoto Prize Commemorative Lectures: Basic Sciences. "Thinking about how living things work." http://www.inamori-f.or.jp/laureates/k24_b_tony/img/lct_e.pdf.

CHAPTER 3

Birth of a Superhero

P53 is the Clark Kent of cancer biology. It spent its juvenile years as a dowdy misfit hanging out on the fringes of the in crowd, and until a decade after its discovery, was pressed into a completely unsuitable profession as an oncogene. It would take a seismic shift in our understanding of cancer to finally reveal its identity as the fabulous cellular superhero, the Guardian of the Genome, subject of well over 66,000 papers, potential anticancer therapeutic target, and number one good guy in a cell's battle against the evil agents of mutation. Throughout all its travails, p53 had one faithful friend, the Lois Lane of this story, who with pleasing symmetry shares a surname, although little else, with the intrepid girl reporter.

Professor Sir David Lane FRS, current scientific supremo of the Ludwig Institute for Cancer Research, past CRUK and A*STAR Chief Scientist, author of several hundred pretty good papers, and pillar of the science establishment, maintains, semi-seriously, that his career has proceeded by a series of lucky accidents. By his account, one might believe that he has stumbled into his present position of scientific eminence in a happy daze, simply because he was performing the right techniques in the right places at the right times. He has even attempted to attribute his being offered an Imperial College lectureship before he had even submitted his PhD to the fact that he was "very cheap to run." However, those who know him well tell a very different story. Asking around his contemporaries, two things emerge about him: one, that he is very tall, and the other, that he is very very smart indeed.

What *is* true is that David's becoming a cancer biologist was not premeditated. He did a PhD in immunology and definitely wanted to be an immunologist. His sideways step into tumour biology resulted not from brains or scientifically motivated curiosity, but from his youthful willingness to handle vast amounts of radioactivity and his desire to get a decent job in the same city as his wife.

David Lane, ca. 1980.
(*Photograph courtesy of Ashley Dunn.*)

David's PhD was at University College London with Avrion Mitchison, one of the last great eccentrics of British science. The son of the novelist Naomi Mitchison and her politician husband Dick, Av came from the fine tradition of scientists who do what they do for the fun of it and don't give a hoot for convention. Unsurprisingly, he was good friends with Jim Watson, who was fascinated by Av's mother, lifestyle, and personality, and Watson was his best man when he married.

Av's style in the lab was, to quote David, "wonderful and awful at the same time." The Tumour Immunology Unit, which Av ran, was funded by ICRF and was one of the hot places to be in the immunology world at that time. In contrast to the rest of UCL, which was very old and fussy, Av's unit was painted in bright colours and was full of bright people to match the décor. However, even Av admits that David was neglected when he started work there, partly because Av was preoccupied with the demands of running a new unit, but partly because that was just what Av did—you were given space, equipment, and a pool of clever people to talk to, and then left to sink or swim. David spent his first year virtually unsupervised, trying to get a horribly difficult assay to work, in the meantime watching the other students and postdocs knocking out *Nature* papers. He was unsurprised, if upset, when Av told him he thought he was no good and should give up, but because it transpired that Av was always firing graduate students and then unfiring them, David survived. He was then plunged back into gloom by the news that Av had hired an American postdoc, Don Silver, to work on exactly the same project, and this time was only prevented from quitting by the fact that he had acquired a new girlfriend, Birgit Muldal. The love of a good woman (whom he ended up marrying in 1975) was enough to keep him in high enough spirits to continue, and it was just as well, because Don, rather than being a competitor, was extremely helpful, ending up as David's de facto PhD supervisor.

Between them, Don and David made a very good team. In between endless conversations regarding life and science, Don taught David to think critically, and also persuaded him to work a little harder (David claims, slightly improbably, to be intrinsically lazy). Don had a background in genetics, and with this added bonus, the pair started to make great progress, publishing a paper in the *Journal of Experimental Medicine*, then the top of the tree as far as immunology was concerned, in 1975. Av was delighted, promoted David to star student status (he now recalls him as "the best student I ever had in University College") (Mitchison Web of Stories), and fixed him up with a postdoctoral position with Bill Paul, a top U.S. immunologist working at the National Institutes of Health (NIH) in Bethesda, Maryland. In passing, Av also sorted out the newlywed Birgit and David's housing issues by installing them in the flat at the top of his rambling old house in Islington, a typically generous gesture. Birgit was finishing up a postdoc at Imperial College and also had a job offer at the NIH, so David just needed to find something to do for the next nine months, and they would then be in sync. All seemed set for a satisfactory transatlantic transfer from one immunology guru to another, followed by a long and happy immunological future.

It was not to be; the world beyond immunology thrust its way into David's life, and neither David nor Birgit ever made it to Bethesda. In David's words, "something happened then that was really strange—rumours got around that I knew how to iodinate proteins. I'd been making my pure protein by setting up a radioimmunoassay for it which was very important for the quantitative immunology we were doing. It was right on the edge—very difficult to do, but it just about worked."

Being able to successfully radioiodinate proteins was, as David indicates, a big deal in those days, and was also a really good way of accidentally dosing yourself with a fair whack of potentially toxic radiation. David's predecessor in Av's lab had, in fact, messed up to such an the extent that he was the Health and Safety Department's poster boy for the "how not to do radioiodination" safety material. The procedure is a way of labeling proteins by attaching hot ^{125}I to them and is very useful for monitoring molecules present at very low concentrations such as the purified protein David was using in his radioimmunoassays; ^{125}I emits γ-rays, so that it is easily detected even when present in tiny amounts. The flip side of this convenience is that γ-rays are so energetic that they are very good at killing cells, which is good if you have cancer, where they are used in radiotherapy, but bad in normal life, where they can not only destroy healthy cells but also mutate them. ^{125}I can be absorbed quite easily whilst doing experiments, mostly because the labeling protocol calls for the radioactive iodine to be produced from sodium iodide and reacted with the protein recipient; the free ^{125}I produced during the reaction is highly volatile and therefore very easy to inhale. Once in the body, ^{125}I migrates to the thyroid gland, where it can profoundly damage the tissue. Av's unit did at least have a fume hood, which gave some degree of protection, but even so, trying not to breathe too much was the preferred option in the crucial parts of the experiment.

The rumours of David's interesting skill set reached Lionel Crawford at the ICRF via Mike Fried's wife, Nancy Hogg, one of David's fellow inmates at the Tumour Immunology Unit. Lionel, as related previously, had realised that the antisera for detecting SV40 T antigens were awful, the assays were, in his words, "cumbersome, irreproducible, and everything else," and that without improving them, nobody was ever going to be able to work out what was going on. Rather than struggling on with what he had, Lionel had had an idea regarding how to quantitate antisera properly but needed a competent immunochemist who wasn't scared of radioiodination to convert his theory into practice. Now, it seemed, that person might exist. He contacted David and asked him to come over to the ICRF for a chat.

Lionel, meticulously well organised both in his thinking and his experimental work, was about as different from the wildly eccentric Av as it was possible to get, but the difference in style was intriguing, the project was interesting, and the salary was seriously good compared with David's PhD stipend of just under £500 a year (equivalent to £4000 in current money). David signed on, two years and nine months into his PhD, without having written up his thesis, which would have to be worked on in his spare time (or more realistically, because he didn't have any spare time, when he should have been sleeping). Lionel was delighted—he had managed to hire someone who was confident with radioimmunoassays, confident with immunising animals, good at the bench, clever, keen, and more than happy to work independently—postdoctoral gold dust.

The ICRF was virtually unknown territory to David and came as a shock to his system. Life in the Tumour Immunology Unit had the speed and ethos of an arthouse movie, with long theoretical discussions during endless coffee breaks, where one's intellectual prowess was perhaps more valued than practical ability. By contrast, the ICRF Fifth Floor was more *Raiders of the Lost Ark* crossed with a hefty dose of *Jaws*. Work was fast, precise, and biochemical, with exciting new data seeming to appear on an almost daily basis, and the predatory sharks cruising the floor did not so much discuss ideas as violently shred them, biting chunks out of their opponents' egos in the process.

It didn't take David long to realise that the inhabitants of this brash new world, although doing great things in many areas, had an immunochemical blind spot. Strangely, because they were rigorously quantitative when it came to the molecular biology techniques that they were pioneering, for immunoprecipitations they were using a bunch of unquantitated antisera made from assorted rodents, with little idea of how good the antisera were. This was all very well when everyone thought they were just looking for one 100-kDa protein, because if that protein showed up, the responsible antiserum was judged a success. Hunting for new SV40 T antigens was a different matter, however; how could you tell whether the many extra bands that always appeared along with the 100-kDa band were specific, or just background? You could get some of the way by comparing immunoprecipitations from tumour-bearing animals with those from tumour-free animals, but even then, the gels run to separate the immunoprecipitated proteins still had so many dodgy-looking regions that assigning one band to one protein was impossible. To quote Lionel: "Once you see things further down the gel, you think 'ooh, what's that?' And the answer is, 'it's a whole mess.'" Alan Smith's lab was trying to get around this problem by translating SV40 proteins in cell-free extracts to get rid of some of the background, but even they hadn't got very far.

The reasons for the awful data were twofold: the crude way in which the antisera were made, and the subsequent immunoprecipitation assay. In those days, antiserum was simply whatever was left over after the blood cells and clotting factors were removed from blood; the resulting clear, yellowish liquid contained a mixed bag of whatever antibodies happened to be circulating at the time of the bleed. Therefore, although it was likely that an animal with SV40-induced tumours would have a lot of SV40-specific antibodies, what the mix of antibodies saw in the tumour and how well they worked to bind antigens could not be guaranteed. As a result, different batches differed wildly in efficacy. Further problems crept in during immunoprecipitation; the antiserum was mixed with a radiolabeled cellular protein extract, and then the small number of antibody–antigen complexes that had formed had to be pulled out from the uninteresting stuff by a variety of methods, all of which were long-winded and inefficient. As a final hurdle, the huge number of other proteins knocking around in the labeled cell extract meant that some of the stickier ones always sneaked through to the end to produce nonspecific background; it was almost impossible to get rid of them by washing the immunoprecipitates because thorough washing tended to destroy the genuine antibody–antigen complexes too.

In the mid-1970s, a new immunoprecipitation protocol appeared that rapidly eclipsed the other methods. Protein A is a surface protein made by the bacterium *Staphylococcus aureus* as part of its defence against being killed when it infects a new host. In its role as point man for *S. aureus*, protein A binds tightly to many antibodies and so traumatizes the cells carrying them so that they either kill themselves or stick their fingers in their metaphorical ears and shout "la la la" whenever they see any new *S. aureus* heading their way (in immunospeak, protein A, a superantigen, elicits apoptosis and tolerance induction).

Because of its ability to bind antibodies, researchers realised that using protein A would be a great way of pulling down immunoprecipitated proteins. Either bound to heat-killed *S. aureus* or subsequently to little beads rather like miniature frog spawn, protein A rapidly attaches itself to antibody–antigen complexes with a tenacity that allows vigorous washing to be deployed to eliminate nonspecific proteins. The results are far cleaner, and the experiments quicker and less fiddly. In David's words: "you just squirt everything in and it works beautifully."

David's first tasks in Lionel's lab were to find a way of testing all of the existing antisera and then to devise a new method of making more specific reagents for the future. Both objectives required him to look at the antibody–antigen complexes formed in immunoprecipitations from a different viewpoint, asking questions regarding the antisera rather than the antigens. To do this, he and Lionel had come up with two rather good ideas: to use radioiodinated protein A as a marker to detect how much complex was present, and to trap and then count the immune complexes on glass fibre filters.

Lionel and David published their paper, "An Immune Complex Assay for SV40 T Antigen," in January 1977, a few months after David's arrival. Their new assay started off as a normal immunoprecipitation, mixing antisera with cell extracts made from one particular line of SV40-transformed cells. Then, instead of spinning them down, the samples were put onto glass fibre filters, which trap immune complexes but not other proteins, and washed very carefully. Protein A labeled by David with ^{125}I was added to the filters, and after some more washing, the amount of radioactivity left on the filters was measured. The better the antibody bound the T antigens in the cell extracts, the more complexes stuck to the filters, and the more ^{125}I was left. The new quantitative tool allowed David to test all of the lab's antisera, bin all the terrible ones, and rank the remainder according to their strength of binding to tumour antigens.

David had achieved his first objective very quickly, and he rapidly realised that it had also given him a way of completing his second task, to make better reagents. However, Lionel was not around as a source of help and advice for this next stage; even before their paper had appeared, he had gone off on a year's sabbatical to Stanford, to work with George Stark, a great mate of his. Fortunately, this event, which would have struck fear into the hearts of many novice postdocs, was so similar to the benevolent neglect David had encountered as a PhD student that it didn't bother him at all: "I didn't expect in any way to be supervised, which is amazing when you think about it now."

Slightly more alarming than being abandoned by his supervisor were the territorial battles after Lionel's departure. Alan Smith and Bob Kamen were both given permanent jobs as proper lab heads whilst Lionel was away, but the jobs didn't come with more space. Possession being nine-tenths of the law, their territorial ambitions meant that David and Peter Piper, Lionel's other remaining postdoc, had to mount a vigorous defence of Lionel's office space and benches. At one point, David himself became the subject of a takeover attempt from the Smith lab, but when Alan came to him with a list of experiments he should perform that week, David just told him that he wasn't interested and carried on with his own projects. Eventually, things settled down, and everyone got back to what really mattered, the research. Joint lab meetings and seminars carried on, and the big open-plan Smith/Kamen/Crawford lab contained some amazingly good people, who were all happy to talk to David: "I had Bob [Kamen] and Richard [Treisman] and Rich Condit on one side, and on the other side was Eva [Paucha] and Alan [Smith] and a vast team including Tony Pawson and Frank McCormick."

Lionel left David a valuable bequest before heading off to California; his technician Alan Robbins, who was very good, very experienced, and as a bonus, was a lot of fun. David and Alan got on famously: "We just treated each other as equals and worked together really well. It was just one of those systems where in the morning you'd do something and in the afternoon the other person would finish it. You knew you could completely trust the other person. The only time it broke down was Friday afternoon. We literally used to *run* to the pub at 5pm on Friday—it was hilarious."

At the end of David and Lionel's paper on quantitating SV40 antisera, there is a beautiful photo of an immunoprecipitation run out on a gel. It is squeaky clean—there are hardly any bands to be seen in the track precipitated using anti SV40 antisera—and, most importantly for David's next step, there was an awful lot of one particular band running two-thirds of the way up the gel: large T antigen. David, in his reading of the immunology literature, had seen a paper in which proteins could be cut out of gels and used to immunise animals to make antibodies, and he realised that if he did the same with large T, he could make antiserum specifically directed against it. This was a big step forward, because all of the previous antisera had been made against whole tumour cells. Having a T-specific antiserum would be like the difference between knowing a postcode and knowing a house number within that postcode. You could go straight to the front door you wanted without bothering the neighbours.

Leaving aside its use as a great immunogen, pure T antigen would also be an invaluable asset for getting going with biochemistry and the new art of protein sequencing. People could study what it was doing and also find the amino acid sequence of its front and back ends; turned back into the DNA three-letter code, the sequence would localise the beginning and end of the large T gene on the viral genome and also match it with the mRNA from which it was made. In retrospect, it isn't surprising that Alan Smith had been keen to suborn David, because this was exactly what his lab wanted to do. The difference, however, was that they were making small amounts of large T in test tubes and would have dearly loved to get their hands on the cellular protein purified by David's technique.

David and Alan Robbins got in a big order of monkey kidney cells from Kit Osborn, Lionel's cell culture technician, infected them with SV40 virus, and made a lot of cell extract. After a massively scaled-up immunoprecipitation, they ran the result out on a big gel, cut out the large T band, solubilised it, and then headed off to the animal house with their precious immunogen.

The story so far, but for the cameo appearances of Kit, Birgit Lane, and Nancy Hogg, has been sadly lacking in female interest. In human terms, this will unfortunately continue, because the Fifth Floor was a very male environment, perhaps accounting for some of the testosterone-fueled antler clashing and similarly daft behaviour that went on there. In the rabbit world, however, a heroine enters at this point; large and white, with big floppy ears and a benevolent expression, Lucy the antibody rabbit single-handedly (-pawedly?) provided David and Alan with the new antiserum they needed. After injection of the large T solution, she produced vast amounts of really good anti-T antiserum whilst remaining in excellent health, unlike some of the other rabbits in her room, who perished in an epidemic of Snuffles, a nasty bunny illness that does exactly what it says on the box. The Snuffles epidemic nearly put paid to the experiment, because the animal technicians wanted to clean out the affected room by culling poor Lucy; fortunately, they relented, meaning that a daring plan to smuggle Lucy out and keep her at home was never executed. Lucy went on to live a productive and reasonably comfortable life (but for the need for regular injections) and produced enough fantastically good antiserum for several years of experiments. She is still remembered with great respect.

David and Alan wrote up their results and submitted a paper in late 1977 presenting their new antiserum. The paper showed that Lucy's anti-T antiserum was incredibly clean and very potent, five times as active as the best hamster antitumour antiserum. The relative size of hamsters and rabbits also meant that she had produced the same amount of antiserum as 500 tumour-bearing hamsters, a considerable saving on all fronts. Most interestingly of all, with the background of nonspecific proteins reduced to almost nothing by David's technical expertise, it was clear that the anti-T antiserum saw not just the large T antigen with which Lucy had been immunised, but also two smaller bands. One of these, the really small one, was SV40 small T, which Lionel, still in Stanford and collaborating with Paul Berg's lab, had recently shown was a splice variant of large T; it made sense that the antiserum should recognise this too because large and small T have a fair bit of common sequence. The other band ran at ~60 kDa; it was not clear what it could be, and David and Alan, instead of speculating, just left it at that, merely saying in the Discussion to the paper that "its relationship to large T and small T is currently under investigation" (Lane and Robbins 1978).

The dry scientific language disguises the fact that before David and Alan's paper, that position on a T antigen immunoprecipitation gel had always corresponded to the awfullest, messiest, region. This was the first time that a single protein had emerged from the confusion. Importantly, because it had been immunoprecipitated with anti-T antiserum, it had to have some relationship with SV40 large T and wasn't just a very persistent background band of no interest. Therefore, the "currently under investigation" of David and Alan's discussion of the 60-kDa band translates as "We are busting a gut to figure out what it is before somebody else does." David was well aware that he would have to really crack on to beat the opposition when it came to identifying it.

In addition to the lab-based excitement, there was a lot else going on in David's life at that time. For a start, he was desperately trying to write up his PhD thesis for submission. In those days, you typed your thesis by hand, referenced it by hand, and made the figures yourself, and David and Birgit spent a lot of nights taking turns to type until their hands hurt in their new house in Chiswick, where they'd recently moved. The other more startling event was that David was now in possession of a rather good extramural job. Av, keeping a fatherly eye on him, had phoned David up one day and suggested that he applied for a lectureship in immunology at Imperial College. Following an interview, David had duly been appointed, despite lacking a doctorate and being only 25. As mentioned before, David claims it was mostly luck, not talent: "I wasn't extraordinary—it was just a set of circumstances and they wanted someone very cheap! At the interview they said, 'We can afford a Rover but not a Rolls Royce' in terms of equipment, so they bought me a couple of things and I set up a little lab there, continuing some of the things from my PhD." It was just as well that David had acquired a motorbike, because after his appointment, he spent an awful lot of time commuting between home, Imperial, and the ICRF (a juggling act that pales into insignificance against his subsequent multijob existence in Singapore, Dundee, and London).

Matters didn't stop there, however, because yet another job offer materialised, once more via Mike Fried, who was inadvertently evolving into David's scientific Fairy Godfather. Having first brought him to Lionel's attention after Nancy Hogg's tipoff, Mike now introduced David to Joe Sambrook from Cold Spring Harbor, who was over on one of his regular romantic/scientific visits to the ICRF. Sambrook liked the look of David and asked him to come to Cold Spring Harbor as a postdoc, to work on hybrid SV40/adenoviruses. Imperial, who had been very relaxed about David spending most of his time at ICRF, seemed equally happy for David to go to America, as long as he showed up to give his lectures and they didn't have to pay him. Birgit got an offer from another lab at Cold Spring Harbor, abandoning her NIH plans, and everything was set, with a start date of late 1978. ICRF did very well out of these arrangements: "I was at ICRF for two years in the lab, but as an employee, probably nine months. I was a very cheap deal for everybody, really!"

Lionel's return from California must have been a bit unpleasant for him. He had spent a happy year doing what he loved best—experiments—and had ended up with some great papers to show for it; as mentioned in the previous chapter, his discovery of how SV40 small T was made was the first ever demonstration of splicing in a coding region. He returned to find that his two postdocs, Bob Kamen and Alan Smith, had not only been transformed into independent lab heads but had been given some of his technicians, and that his own lab had consequently shrunk to almost nothing. Bob remembers him being remarkably generous about it: "Alan and Lionel and I got together and I thought it would be controversial but Lionel said 'Fine, let's split it three ways.' So I got the left two bays, Lionel got the middle two bays and Alan got the right two bays. It was hard to tell if he minded—he was very, very British, non-confrontational. I don't think he minded. I was certainly very happy he hadn't created a problem."

David had also changed: "When Lionel came back, he met a very different creature. He met someone who'd got an independent position and written a rather nice paper in a journal and had discovered this other protein, and hadn't put his [Lionel's] name on anything! I didn't know, because Av hadn't taught me, how things worked in the rest of the world. The work with Alan, I just told Lionel the paper had been accepted by *Virology*, and wasn't that good? When he said he was surprised I hadn't put his name on the paper, I was truly astonished! Looking back, I must have seemed very arrogant, but it was more naïvete than anything."

Leaving aside the charged personal situation, both Lionel and David were very excited by David's mystery 60-kDa protein. In June 1978, David got his PhD, finally justifying the "Dr. David Lane" sign on his Imperial office door, and was able to really turn his mind to what could be going on.

In the first half of 1978, David had spent a lot of time tinkering with the fine detail of his immunoprecipitations, as had Lionel, who, somewhat to David's chagrin, was mirroring the work to get the gels looking better. There were several items that could be tuned up, and all were attended to. David realised that preparing the cell extracts by gently popping the cells open with a smidge of detergent rather than the accepted method of plunging them into dry ice repeatedly meant that the proteins inside didn't fall apart. Intact protein meant fewer bands on gels, and more material running at the right size, and was especially important for large T, because it had a nasty tendency to disintegrate into a set of bands running around 60 kDa, confusing any analysis of that region. The antiserum, now that it had been properly tested for activity, could be used in minimal amounts (1 µL—a thousandth of a millilitre—of Lucy's antiserum was enough to get a good clean immunoprecipitation), and pulling antibody–antigen complexes from the extracts using protein A resulted in even more reduction of background.

Having optimised the methodology, David then started comparing cell lines for amounts of 60-kDa protein. As he'd observed previously, untransformed cells had no detectable 60-kDa protein at all, and, in fact, the cells in which he'd first seen it, the SV40-infected monkey kidney cells, didn't contain very much either. Far better was an SV40-transformed line, SVA31E7, which had boatloads of it. With more protein visible, it was also easier to estimate size; comparison with a good marker ladder (a mix of proteins of known molecular weights, run beside samples) showed that a more accurate weight estimate for the mystery protein was 53 kDa. It was duly christened 53K, which was a bit vague, but would do until someone worked out what it did and renamed it accordingly.

The big question with the 53K protein pertained to its relationship with large T. Was it showing up on the immunoprecipitations because the relationship was familial, as for small T? Both the large T and 53K bands were always there following immunoprecipitation by lots of different antisera, either Lucy's rabbit anti-T or the old antitumour stuff, and each protein was equally well recognised, but this didn't necessarily mean that they were related. The killer experiment would be to do an immunoprecipitation, run the samples on a gel, cut out the large T and 53K bands, and then see whether the isolated bands could still be recognised by Lucy's anti-T serum in an ^{125}I filter-binding assay.

David did a spot of breath-holding and involuntary self-iodination to make some more radiolabeled protein A, and the experiment went ahead. It had a very unexpected result; the isolated large T band was happily recognised by Lucy's antiserum, but the 53K protein was not seen at all, despite numerous attempts to prove the contrary. The conclusion had to be that the 53K protein did not have any of the same antigenic determinants as large T and wasn't related to it at all. So what on earth was it?

The most obvious answer, that it was an artefact and the 53K protein was just sticky rubbish, could be quickly excluded in David and Lionel's minds; they knew that their extracts and gels were irreproachably good. This left only one possibility, that the relationship between large T and the 53K protein was not familial, but conjugal; the 53K protein, like an overpossessive spouse, was so clingy that in anti-T immunoprecipitations it always came along for the ride. In fact, it was bound so tightly that only separation on a gel could tear it away.

So far, so good, but this still didn't solve the question of where the 53K protein was coming from. Because it only showed up in SV40-transformed cells, logic dictated that it ought to be made by the SV40 early region, but this was becoming increasingly unlikely; that year, the SV40 DNA sequence was coming out in dribs and drabs, and it showed that there was simply no room in the viral early region for another completely different protein in addition to large and small T. Something else, something novel, had to be going on.

Lionel and David suddenly realised that what they were looking at might be partial vindication of the years of pouring millions of pounds and dollars into research on small DNA tumour viruses. The whole point had been to use the viruses as a way of finding the cellular control mechanisms they were subverting to cause cancer, and it followed that to subvert the control mechanisms, contact had to be made with the cellular control proteins. That was it! The 53K protein must be *cellular*, not viral. Large T, far from establishing a conjugal relationship with it, had forced the transformed cell to switch the 53K protein on and was holding it hostage.

Thomas Edison famously said that genius was 1% inspiration and 99% perspiration. Whatever the ratio for David and Lionel in the matter of the 53K protein, once they realised that large T was contacting the cell via the 53K protein, it was as if one of Edison's patent incandescent light bulbs had come on in their heads. An interaction that tight had to be important for transformation, although how, exactly, was not a question they would be able to answer any time soon. What was most pressing was to make the case for the cellular origin of the 53K protein as watertight as possible, and publish. Their technical expertise meant that they had an edge over the competition, but they couldn't rely on that forever.

First up was to test all the other antitumour antisera in the filter binding assay, to see if they behaved similarly to Lucy's anti-T serum. Surprisingly, although most of the antisera were identical in their inability to recognise the gel-isolated 53K protein, one antitumour serum bucked the trend, still seeing both proteins after gel purification. The conclusion was that in that particular tumour, the mouse host had made antibodies not only to T antigen proteins, but also to the cellular 53K protein. This had to be properly proven, so David did a collaborative experiment with Eva Paucha, Alan Smith's postdoc (extremely nice, very talented, and very beautiful, Eva Paucha was Alan's partner not only in the lab, but outside it; her early death a decade later at the age of 38 was a huge tragedy). Using the Smith lab's cell-free translation system, Eva showed that there were conditions under which the 53K protein could be detected by the oddball antimouse serum even when it wasn't complexed to large T.

Having the strange mouse antiserum allowed David and Lionel to do one final very important experiment. If, as they thought, the 53K protein was important for transformation and cancer, then it should show up not just in SV40-induced tumours, but in other sorts too. David did one final filter binding assay, using protein extract made from polyoma-transformed cells, which had never had so much as a sniff of SV40. All of the anti-SV40 antisera failed to bind anything, except for the strange antiserum, which saw a protein near enough in size to the 53K protein to suggest that it was very closely related, if not the same item.

David and Lionel wrote up their data in a short, three-figure paper entitled, "T Antigen Is Bound to a Host Protein in SV40-Transformed Cells," sent it to *Nature* in November 1978, and sat back, fingers crossed. If the reviewers liked it, it would with any luck be published before the turn of the year, thus forestalling any arguments about who had got there first. Sadly, the reviewers did not like it at all; it came back with all sorts of objections attached to it, most of which were also voiced by other members of the SV40 community when Lionel sent them preprint copies (the generosity of those days, when even big results were shared with other labs before publication rather than hugged to one's chest for as long as possible, is astonishing to the modern eye). The main complaints seemed to be that (1) nobody believed it; (2) despite that, everyone had done it already; and (more coherently) (3) there was no proof that a complex had formed other than the immunoprecipitation data, and something else was needed. Because the last point was something tangible that David and Lionel could address, they did a hasty biophysics experiment that brought the number of figures in the paper up to four. David's opinion of this "reviewers' experiment" is unflattering to all concerned: "The critical thing that we did that satisfied them, although it was a crap experiment, was to run some gel filtration columns and show that they [large T and the 53K protein] came out together. But if you look at the figure, it's pretty awful!" It was enough, however, and *Nature* accepted the paper at the end of January 1979.

By this time, David had moved to Joe Sambrook's lab at Cold Spring Harbor; thus, the final revisions were completed by transatlantic phone call and the proofs corrected on a flying visit home, in Lionel's house in Highgate. The paper came out on 15 March. Unfortunately, by this time, the competition had caught up, and so officially, David and Lionel published the discovery of the 53K protein in the same year as three other groups, rather than the year previously.

The speed with which the other labs got their papers out is very illuminating. As had happened to Bob Kamen during the discovery of splicing, other people had got similar stuff but hadn't hit on the right explanation for what was going on (gallingly for the authors involved, three papers published in 1978 identified a band corresponding to the 53K protein but failed to realise its significance). As Lionel says, "so many of these things, you just think 'ah, *yes*!' Several people had the data and once they were provided with an explanation were able to write up very quickly."

Two more papers, by Daniel Linzer and Arnie Levine at Princeton, and from Pierre May's group in Villejuif, also looked at the 53K protein from the SV40 angle and confirmed David and Lionel's data. Linzer and Levine additionally found a protein that was similar or identical to the 53K protein in embryonal carcinoma cells. This result was important because it showed very clearly that the 53K protein was present in a cancerous cell type that had nothing to do with SV40 or polyoma, implying that it might have a more general role during transformation. May's group published more of a roundup of all of their immunoprecipitation data, with no speculation as to what might be going on beyond the observation that some of the bands were potentially of cellular origin. What their paper *does* illustrate very well is the sheer number of bands that it is possible to immunoprecipitate with anti-SV40 tumour antisera—it is small wonder that the authors were unable to work out which might be of greatest importance.

The other paper coming out that year, from Albert DeLeo and Lloyd Old from the Memorial Sloan-Kettering Cancer Center in New York (together with colleagues from the National Cancer Institute in Bethesda), was crucial in establishing the wider relevance of the 53K protein, showing that Linzer and Levine's embryonal carcinoma data were not just a one-off. DeLeo's paper provided the link to other forms of cancer and showed the protein was not just a virally induced phenomenon. In it, the investigators identified a tumour antigen of cellular origin that was never present in normal cells but always showed up in transformed cells, irrespective of the carcinogenic insult. Every transformant they tested, whether spontaneous or chemically or virally induced, contained the antigen, which they called p53.

Their competitors' interest in David and Lionel's data was not mirrored with much enthusiasm by the wider DNA tumour virus field. Even Fifth Floor reactions to the paper were muted. The Smith lab published a beautiful *Cell* paper subsequently in 1979 that established conclusively that the 53K protein was of cellular origin and had a completely different protein sequence from that of large T, and they continued working on the problem after Alan's departure for the National Institute for Medical Research, but in the polyoma-oriented labs, the reaction ranged from Not Interesting: "I was anti—I just thought it was another host binding protein" (Mike Fried); "I didn't really focus on it" (Bob Kamen), to Possibly Interesting: "Because it was a band that wouldn't go away when subject to real stringency, there was a sense that this might be something" (Adrian Hayday). Outside the ICRF, in his new home of Cold Spring Harbor, David encountered outright hostility. "[One of the senior staff investigators] came up to me and said they didn't understand how I could publish something like that, as it was obviously wrong. I had a complete meltdown! People didn't accept the immunological idea—they thought that for a protein to bind that tightly so that it could resist washing, processing and all that, was just not possible." David, with his immunological background, knew that there was established precedent for such tight protein–protein interactions in the immune system, but sadly, his less well-informed critics remained unconvinced.

Immunological ignorance aside, the main reason for the underwhelming response to the 53K protein was that by the late 1970s, DNA tumour viruses were widely regarded as the best tools for dissecting the mechanisms of eukaryotic transcription, translation, and DNA replication, but had fallen out of fashion in terms of cancer research. The big hitters in the molecular biology of cancer were now the RNA tumour viruses, thanks to Harold Varmus, Dominique Stehelin, and Mike Bishop's 1976 discovery of the nature of retroviral oncogenes.

In a paper that won Varmus and Bishop the 1989 Nobel Prize for Physiology or Medicine, the v-*src* oncogene of Rous sarcoma virus was unmasked as a cellular gene gone bad; Rous sarcoma virus had picked up a copy of an important cellular control gene, c-*src*, and mutated it into a form that caused unregulated cell proliferation, and hence cancer. Researchers everywhere leapt enthusiastically onto the retroviral oncogene bandwagon, working on what c-*src* and v-*src* might be doing, and also trying to identify further captured cellular oncogenes in their own pet retroviruses (e.g., at the ICRF, Tom Curran, a PhD student with Natalie Teich, found the v-*fos* oncogene in the FBJ murine osteosarcoma virus). This flurry of activity rapidly confirmed that the capture and corruption of cellular genes, so-called proto-oncogenes, had happened multiple times, and that the genes in question played important roles in both normal and cancerous development.

The balance tipped even further in favour of RNA tumour virus research in 1981, when Bill Hayward, at the Rockefeller Institute in New York, showed how Avian leukosis virus, ALV, causes cancer. ALV belongs to a second group of retroviruses that do not carry their own personal oncogene, but are still extremely tumourigenic because of a peculiarity of the retroviral lifestyle; after infecting a cell, all retroviruses make a DNA "provirus" copy of themselves using reverse transcriptase and then insert the provirus into the host cell's DNA, where it becomes a passenger in the genome. When Hayward examined where in the genome his ALV provirus had put itself, he found it next door to a growth-controlling gene called c-*myc*, which had already been identified as the cellular proto-oncogene kidnapped by another virus, MC29. Hayward showed that powerful transcriptional control elements in the ALV provirus were forcing the cellular c-*myc* gene to make inappropriately high amounts of c-Myc protein, driving cells into constant proliferation and causing a tumour to develop. He christened this new method of tumourigenesis "insertional mutagenesis."

RNA tumour virus research really seemed to have done its job and cracked the cancer problem; with the retroviruses, Mother Nature handed researchers the means of identifying scores of cellular genes important for regulating growth. Although it would not be trivial to work out how these genes worked, it was as though the doors had suddenly been opened into a hitherto locked building, and light was shining into many unexplored rooms.

In this new dawn, the DNA tumour viruses began to look almost irrelevant. Their oncogenes did not have cellular counterparts. They did not cause insertional mutagenesis—they seemed to integrate randomly in the genome. The 53K protein was one of only a few established links to the cell, and it was, frankly, a boring band on a gel, with nothing to recommend it other than its insistence in showing up in a lot of transformed cell lines. Figuring out what it did would require it to be cloned, and cloning cellular genes was incredibly hard. Far better to work on retroviruses, with their dinky little genomes that made cloning individual genes comparatively easy.

Research on the 53K protein did continue, but only in the small number of labs with an existing interest; new recruits were very thin on the ground. The smallness of the field did mean that one problem dogging the protein could be resolved relatively harmoniously—the question of what to call it. Everyone who had published on it had come up with a different name—the 53K protein, Non Viral T Antigen, the 54K protein, p53, phosphoprotein53. After some arguing, the field agreed on "p53," and thus it has appeared in all subsequent literature. Ironically, p53 is not 53 kDa in size; the protein has some oddities that make it run abnormally slowly on gels, and its weight is actually ~43 kDa or less. However, by the time people figured this out, the name had stuck.

The early work on p53 had a dual focus. Firstly, most labs harboured at least one unfortunate soul stretching the existing technology and their own ingenuity in the race to clone the p53 gene, a feat accomplished in 1982 by Peter Chumakov and colleagues in Moscow, closely followed by Moshe Oren in the Levine lab, and Lionel's group. Secondly, efforts to characterise the p53 protein were ramped up, made easier by great improvements in the antibodies against SV40 large T and the 53K protein. A new, powerful immunochemistry reagent, the monoclonal antibody, had appeared in 1975, when George Köhler and César Milstein at the Laboratory of Molecular Biology in Cambridge fused antibody-producing B cells and myeloma cells to make hybridoma cell lines that grew in culture indefinitely and made just one antibody type, directed against a single antigen on a single protein. To resume the postcode analogy of antibody binding, if antitumour antisera see a postcode, and anti-T antisera a specific house within that postcode, a monoclonal antibody can go into the house and find the toaster in the kitchen. The ability to achieve such exquisite specificity in antigen recognition has revolutionised both experimental science and medicine, where monoclonal antibodies are widely used as "magic bullet"–style therapeutics. Not surprisingly, their invention garnered Köhler and Milstein a Nobel Prize, in 1984.

David, at Cold Spring Harbor, and Ed Harlow, a new PhD student in Lionel's lab at the ICRF, made some of the earliest and best monoclonals against SV40 large T and p53. David was having an interesting time in Sambrook's lab: "When I went to ICRF I thought I'd landed on Mars, but when I went to Cold Spring Harbor it was *definitely* Mars! Joe Sambrook was very ruthless. He had two types of postdocs, regular American ones, who were properly selected and carefully interviewed and came in to work on particular projects, and then he had a bunch of bright English people he threw in there. If you weren't any good you just left. There was no equipment. There was nothing, so people would steal. It was chaos—disorder, total selfishness and absolute obsession. It's kind of miraculous what we all managed to do there."

David and Ed Harlow had overlapped briefly in Lionel's lab and hit it off instantly. This was unsurprising because, to be honest, Ed's brand of laid-back friendliness is hard for anyone to resist. Born in Southern California and ending up in Tulsa, Oklahoma after a peripatetic childhood, Ed decided he wanted to see the world and wrote to Lionel for a job. He pumped gas for six months to make enough money to come and arrived entirely grantless, clad in Oshkosh dungarees, and equipped with

Ed Harlow orders lunch, ca. 1992.
(Photograph courtesy of author.)

nothing but intelligence, ambition, and a willingness to work (and party) very hard. Lionel, entirely charmed, found him the money to stay for a PhD. It quickly became evident that beneath his happy-go-lucky exterior, Ed was the real deal: a thinking, original scientist who was also incredibly good at the bench. His project consisted of making monoclonals and plugging them into the SV40 system, and he published eight papers during his three years at the ICRF. With Frank McCormick, a postdoc with Alan Smith and yet another scientific star in the making (a pioneer of the biochemical dissection of cancer and of targeted cancer therapy, Frank became Director of the UCSF Cancer Center in San Francisco), Ed did the reverse experiment to Lane and Crawford, showing that a monoclonal antibody against p53 would coprecipitate large T antigen, and getting himself a *Nature* paper in the process.

Now a Harvard Professor of many years' standing, and a Senior Advisor to Harold Varmus, Director of the U.S. National Cancer Institute (and an old friend from his ICRF days), Ed's memories of the ICRF are what everyone should get from a good PhD experience: he had a great time, found a lot of friends, and began to realise he could make it as a scientist:

> ICRF was a breeding ground for a large number of folks that have gone on to do exceedingly well. I don't understand what made this group special, other than the good science at ICRF attracting committed individuals who did well over time. One thing that might be true is that we did get pushed and subsequently learned that we could do useful things. That confidence was built at least in part from the environment. In a productive but somewhat perverted way, the ICRF 5th floor lab heads were great models. Their allegiances to and disputes with one another changed over time, but their commitment to science was always first. I learned that one didn't have to be deep personal friends to respect the quality of science that was being produced. [However], I am now very committed to not doing science with folks I don't like! A simple rule that I have adhered to for years is that if I don't want to have a drink with someone, I don't want to work with them.
>
> In many ways, ICRF and London were a rite of passage—I continue to view London as the most important stage of my career. I came from a spirited but second-class training, and I learned what world class science was at ICRF: important questions, rigorous answers, and the excitement of discovery. I learned that I thrived in big cities, that having access to first class cultural events—particularly art and theatre—was important to me, that I wanted to work in the best scientific environments, and that I could compete at that level. The scientists from my ICRF days have become an

important circle of friends whom I continue to cherish. I would go to huge lengths to help them if they needed it, and I'd be surprised if they wouldn't reciprocate if I called on them for any type of favour.

Ed's evident star quality made him an inevitable target for recruitment to Cold Spring Harbor, and he went off there in 1983 to work on the adenovirus E1A oncogene, leaving behind a case of Dubonnet for Kit Osborn in thanks for her tissue culture virtuosity. No one was more pleased for him than David: "Ed's an amazing guy, a very generous person and easy to get along with. He was very excited about monoclonal antibodies and he got more and made better ones than me, although it was OK as we were very friendly and we swapped the reagents around a lot. Cold Spring Harbor realised they needed somebody who could make monoclonal antibodies and I recommended him for the job. It worked out incredibly well." Despite David having left Cold Spring Harbor to finally take up his lectureship at Imperial College by the time Ed arrived, the two men stayed close, ending up coauthoring a best-selling methods handbook *Antibodies: A Laboratory Manual*, whose reassuringly calm green covers enclose an idiot's guide to working with proteins, has sold more than 40,000 copies, and has smoothed the paths of numerous graduate students and postdocs with its easy-to-follow recipes and sensible advice.

Back in the world of p53, although a dribble of papers characterising where and when the protein could be found were published between 1979 and 1982, until the p53 gene was cloned no clues to its function were forthcoming. One of the most significant papers in this frustrating interim period was from Lionel's lab, in 1982. p53 had already been shown to crop up in many tumour cell lines and never appeared in normal cells (with the exception of some early embryonic types), but Lionel, his technician David Pim, and Richard Bulbrook, head of the clinical endocrinology lab at ICRF, took this a stage further, by examining real cancer samples. They looked at p53 levels in antisera taken from breast cancer patients and normal controls, and found that whereas p53 was undetectable in the controls, 9% of breast cancer samples tested positive. The paper is the first demonstration of the link between p53 and human cancer, but what exactly that link was would take a further seven years to be revealed.

1983 saw a sea change in p53 research. With the gene cloned, it was now possible to introduce it into tissue culture cells and see what it did. The prevailing dogma from the retroviral oncogene jocks was that cellular genes could be turned into oncogenes simply by switching them on at high, constant levels. The p53 gene was therefore overexpressed in tissue culture, and sure enough, it looked like a pretty decent oncogene. Work from multiple labs showed that it would cooperate with another oncogene, *ras*, to transform primary cells, that it could immortalise cells if added in by itself, and that it was really good at souping up established but tumourigenically weedy cell lines to make them far better at causing tumours in animal models. p53 was also shown to bind to the adenovirus E1B oncogene by Arnie Levine's and Alex van der Eb's labs, establishing it as a point of convergent evolution for DNA tumour viruses.

Despite having made it into the oncogenic den, p53 was not home and dry. It started to become obvious that, like a sheep in wolf's clothing, p53 was not behaving with the savagery expected of a bona fide oncogene. Very early on, Varda Rotter from the Weizmann Institute started seeing strange anomalies between her data and the party line. Adrian Hayday recalls: "I remember being at a Cold Spring Harbor conference in ~1983 when Varda Rotter did a 10 minute talk presenting data saying all her tumours had p53 disrupted, so how could p53 be an oncogene? Everybody got up and started explaining how that couldn't be right, that although they didn't doubt her data, p53 must be activated in some other way." Rotter continued to find that p53 seemed to be lost in certain cell lines, and in 1985 she was backed up by Sam Benchimol (an ex-Crawford postdoc) and Alan Bernstein, who showed that p53 was inactivated in some retrovirally induced tumours. Clearly, something very odd was going on; in some labs, p53 was definitely an oncogene, and in others, it couldn't possibly be.

The solution to the puzzle, when it came in 1988 from the labs of Arnie Levine, Varda Rotter, and Moshe Oren, was a bit embarrassing. Comparison of the many p53 clones knocking around showed a marked schizophrenia in oncogenic potency; some worked very well in transformation assays, and others didn't work at all. Instead of blaming the experiment fairy for the failures in transformation, as had been happening for some years, people started to look properly at the DNA sequences of all of the different clones. It transpired that the field had rather carelessly been working with two versions of p53: a mutated oncogenic one, which had been cloned from transformed cell lines, and a normal, wild-type copy, cloned from primary, untransformed cells. Furthermore, it wasn't just a fluke that all of the mutant p53 clones had come from transformed cells; in 1989, Levine and Oren's labs both published that the wild-type p53 proto-oncogene could actively inhibit transformation, if added to cells in combination with genuine oncogenes. It appeared that mutation of normal p53 was actually a prerequisite for transformation, and, by extrapolation, cancer.

So what was going on? The answer was already out there but had been ignored for many years: p53 was a tumour-suppressor gene.

Scientists should not display a herd mentality, and definitely should not gallumph off in a particular direction merely because it is trendy, but in the 1980s, the torrent of important new information pouring out regarding retroviral oncogenes and their cellular counterparts was extremely compelling. The widespread assumption that finding cellular oncogenes would solve the mystery of cancer managed to obscure an underlying fact regarding cancer that everyone knew, but almost everyone chose to forget: Cancer is a rarity, a one-in-a billion event where a cell's normally tight grip on its growth potential goes badly wrong. Furthermore, it is usually a slow disease, a disease of aging. The chance of getting cancer increases roughly a thousand-fold between the ages of 30 and 80, and even when there is a known carcinogenic insult, such as exposure to radiation, any resulting cancer takes years to develop.

Because of the dominance of viral oncogenes, research had been heavily skewed towards asking what caused cancer, rather than the equally important question of what stopped it. How was it that a powerful oncogene that could transform an immortalised cell at a single stroke was so inefficient under most circumstances? The answer, obviously, was that cells had to contain genes capable of combatting the effects of oncogenes. Such genes, variously called antioncogenes or tumour-suppressor genes, would have to be inactivated in some way in order for an oncogene to be able to exert its effects. p53 fitted the bill exactly.

Presciently, the notion that p53 was a negative regulator of transformation had been suggested by David and Lionel in their original 1979 paper, although their theory vanished under the combined weight of the field's expectations that p53 would be an oncogene. Lane and Crawford proposed that p53 "might normally act as a regulator of certain cellular functions related to growth control and itself be neutralised by binding to T antigen." In other words, p53 was the handbrake stopping the cell from behaving inappropriately, and large T, by releasing the handbrake, unleashed mayhem.

The idea of such a "handbrake" protein was not a novelty, at least not to those in the DNA tumour virus field whose roots were in prokaryotic molecular biology. In 1959, François Jacob and Jacques Monod came up with the concept of the repressor—a gene product that could act to switch off other genes, and in the years that followed, many prokaryotic examples of such proteins were discovered. It seemed perfectly reasonable to Lionel that animal viruses would also have evolved such a mechanism; if you have a tiny genome, you must travel light: "You have to think how it feels from the virus's point of view. It's cheaper to inactivate a control, as it's already there, you don't have to make anything. It's as if you've got your handbag, and all you do is undo the clip to get at the contents. It's more efficient, it's cheaper and it's faster." To David, coming from a different tradition and also free of preconceptions, the logic of the argument seemed clear too: "I remember thinking a lot about it—that paper was quite carefully written. I guess all I thought about was how the oncogene affects the cell."

Luckily for the p53 field, the news that their protein was, in fact, exactly the opposite of what they thought it was came at exactly the right time. In 1989, tumour suppressors were right back in fashion, and furthermore, in one of the biggest scientific splashes of the previous summer, their relevance to DNA tumour viruses had been shown by someone well acquainted with the travails of p53: Ed Harlow.

Ed's work on the adenovirus E1A oncogene at Cold Spring Harbor had taken him down a similar path to his PhD project on p53 and large T. The early region of adenovirus, like SV40 and polyoma, contains genes necessary to drive the host's replicative machinery, which can double as oncogenes. One of these, E1B, had already been shown to bind p53. Ed, in similar experiments to those he'd performed so successfully as a student, was looking at what E1A, the other oncogene, bound. His lab established that it saw a number of cellular proteins, and that one, 105 kDa in size, was present in large enough amounts in a cell that it would be worth trying to purify. Protein purification is a real slog, so in the long days and nights of growing cells and running columns, the Harlow lab took up an alternative strategy—reading the literature to see if anyone had cloned an interesting 105-kDa protein already. It paid off in spades. In 1987, Ed was temporarily back at ICRF, working on *Antibodies: A Laboratory Manual* with David (who had by this time moved back from Imperial to ICRF Clare Hall, as related in the next chapter), when he saw in the library a paper by Wen-Hwa Lee and colleagues, reporting on the protein encoded by the retinoblastoma gene.

Retinoblastoma is predominantly a cancer of childhood; most cases are diagnosed in children under the age of five. If caught early enough, it has a 90% cure rate, but there are longer-term risks associated with recurrence of cancer and also with the side effects of treatment. Blindness and reduced vision are common. How retinoblastomas develop had long been a subject of interest to scientists, because it was very clear that there was a hereditary component; there were retinoblastoma families.

In 1971, Alfred Knudson used retinoblastoma as the model for his famous "two-hit" hypothesis. After doing a statistical analysis of patient data, he proposed that if a baby was born with an inherited ("germline") mutation in the gene responsible for retinoblastoma, there would only have to be one further mutation, in the other healthy copy (allele) of the gene, for cancer to arise. Such babies developed multifocal cancers of both eyes as very young children. In contrast, if the first mutation was not inherited, but happened spontaneously, it would take longer, but a further hit to the other allele of the gene would again cause retinoblastoma, although it would be in just one place in one eye, and at a later age. Crucially, for the two-hit hypothesis to be true for retinoblastoma, the gene causing it had to be a tumour suppressor; complete inactivation caused by loss of both copies was required.

The race to clone the retinoblastoma, or Rb, gene, was hotly contested in the nascent field of human genetics. Bob Weinberg's lab, publishing in 1986, got a partial gene, but in spring 1987, Wen-Hwa Lee and colleagues published the complete Rb gene sequence, following it up with a paper in October of the same year describing the Rb protein. It was this latter paper, with its demonstration that Rb was nuclear, ~105 kDa in size, and likely to be involved in gene regulation, that was the answer to Ed's prayers.

Lab heads always hope that in their absence, their labs will not turn into empty wastelands covered in "Gone Fishing" signs, so Ed, on his return to Cold Spring Harbor a week later, was gratified to find that two members of his lab, Karen Buchkovich and Peter Whyte, had not only seen the Lee paper but had already started on the key experiments. Because it was just a case of slotting the new protein into the well-oiled expert machinery of the lab, data came very quickly. In a miracle of concentrated experimental toil and journal editors' fast-tracking, "Association between an Oncogene and an Antioncogene: The Adenovirus E1A Proteins Bind to the Retinoblastoma Gene Product" was published in *Nature* in July 1988. The Harlow lab's paper showed that in addition to mutating the Rb gene itself, as occurred in retinoblastoma, there was another way of inactivating it that also led to cancer; another protein, in this case E1A, could bind to it and block its activity. This latter result, that a DNA tumour virus could cause cancer by binding and inactivating a tumour repressor, was entirely novel, totally different from the way retroviruses worked, and a huge conceptual breakthrough for the field.

To David, watching an excited Ed jumping around the Clare Hall lab waving a copy of Wen-Hwa Lee's paper, it was a small step to realising that all of the anomalous p53 data could also be explained if p53, like Rb, was a tumour suppressor gene. However, this was one scientific race he didn't win: "I can remember telling Ed I thought p53 was a tumour suppressor like Rb, that it was the same, and I remember seeing how the sequences had all these mistakes in, but I didn't react quickly enough."

Although the Levine and Oren labs had shown that wild-type p53 could block transformation in tissue culture, this was not enough to result automatically in p53's enrollment in the tumour-suppressor club: to be a true tumour suppressor, p53 had to be mutated in real human tumours; it had to cause an increased incidence of cancer in people with germline p53 mutations, and knocking it out in experimental animals should result in an increased level of tumour formation. All of these criteria were fulfilled in short order. In 1989, in a ground-breaking paper, Bert Vogelstein's lab showed that p53 mutations and deletions occurred in more than 75% of human colon cancers, and it soon became apparent that loss or mutation of p53 was rife in human cancer in general. Today, it is estimated that at least half of all cancers have defects in p53, and of the rest, most have related defects that stop p53 working properly. In 1990, germline mutation of p53 was shown to be the cause of Li-Fraumeni Syndrome by the labs of Steve Friend and Esther Chang. Li-Fraumeni sufferers have a hereditary predisposition to cancer and develop multiple primary tumours from a very young age, with a 50% risk of developing cancer by age 35, and a 90% lifetime risk; it is as well that the syndrome is extremely rare. Finally, in 1992, Allan Bradley's lab knocked out the p53 gene in mice and showed that the resulting p53-null animals were developmentally normal, but, just like Li-Fraumeni patients, had a vastly increased tendency to develop tumours from a very young age.

❖ ❖ ❖

As David and Lionel had realised more than a decade previously, techniques for looking at protein–protein interactions are dependent on good antibodies, and the battery of antibodies and antisera developed for the DNA tumour virus proteins, Rb and p53, were now very good, indeed. It was therefore unsurprising that many loose ends were wrapped up in the DNA tumour virus field in a very short time, by the simple expedient of seeing to which viral proteins p53 and Rb were able to bind. The oncogenes of SV40, polyoma, adenovirus, and the wart-causing papillomaviruses were all shown to bind both p53 and Rb. When studies on Rb, many from Ed's lab, showed that it was a major regulator of the cell cycle, the last piece of the puzzle dropped into place. Although some viruses, like polyoma, had some additional oncogenic weapons able to activate cellular growth processes, it was the viruses' interactions with Rb and p53 that laid the foundations for cancer. Inhibition of Rb forced the cells into cycle, and, thanks to p53, once they were in cycle they were unable to stop. Normally, in a lytic infection, the cell would be killed when the viruses burst out of it, and no long-term harm was done, but on the rare occasion that a virus inserted itself into the host cell genome, it continued to send out signals for unregulated growth, and these led, eventually, to a tumour.

Forty years after Max Delbrück decided animal viruses might be interesting, DNA and RNA tumour viruses had more than exceeded expectations and completely vindicated the reductionist philosophies of the phage group and their successors. By studying these small viruses, researchers had laid open many of the mysteries of cells: how they replicated; how they transcribed and translated their genes into proteins, and most importantly, how they handled information. The proto-oncogenes and tumour suppressors discovered thanks to tumour virus research really were the way in to the complex networks of proteins whose interactions govern how a cell responds to all eventualities, from its birth to its death.

p53's solo career has taken off beyond all expectations from its days as a humble unknown band on a gel. Working out which biological processes the p53 protein affects and the molecular mechanisms it uses to protect cells from cancer has occupied legions of researchers in the 20-odd years since Vogelstein's seminal paper. Although the molecular details grow ever more complicated and are far beyond this chapter's scope, the consequences to a cell of p53's actions are fairly clear, and have been almost from the start. In 1992, David Lane wrote a short review article in *Nature* summarising the field's discoveries to date. The article, cited more than 3000 times, cut a channel of clarity through the intimidating morass of p53 data, and in addition coined the snappiest-ever title for a gene: p53, the Guardian of the Genome. p53's special superhero name caused a great deal of envy amongst some retinoblastoma devotees, who tried very hard to think of an equivalent, but never quite managed it. Captain Cell Cycle? The Tumour Terminator? The Policeman of Ploidy? None really has the same indefinable glamour and star quality.

So, what does the Guardian of the Genome do? In his article, David proposed the model that still stands today, that normal p53 "acts as a molecular policeman, monitoring the integrity of the genome" (Lane 1992). p53, amongst other things, is a transcription factor and can directly regulate multiple genes by switching them on or off as required. It is normally present at very low levels in a cell, but if it detects any DNA damage, whether it be caused by irradiation, chemical agents, or a rogue virus trying to sneakily replicate, p53 builds up to far higher concentrations and triggers multiple alarm systems. Depending on the circumstances, it then makes a decision about what the compromised cell should do next. If the DNA damage is fairly minor, p53 stops the cell from cycling, calls in the repair gangs, and gets the damage fixed, after which the cell can proceed on its way. If the damage is bad, p53 either triggers a state called senescence, where cells doze off, never to cycle again, or pushes the destruct button, writing off the cell as injured beyond repair.

From this, it is clear why p53 loss or mutation is so common in tumours; once it is gone, all hell breaks loose in the genome. If its genome is damaged and cannot be repaired, a cell starts to build up mistakes in its chromosomes every time it replicates. The mistakes are sometimes so catastrophic that the cell dies anyway, but all too often, they result in the loss of other tumour-suppressor genes and the conversion of proto-oncogenes into frank oncogenes, leading to unrestricted growth, and cancer.

David Lane, 2002.
(*Photograph courtesy of CRUK London Research Institute Archives.*)

p53 is therefore unimportant during a cell's normal life, yet crucial under battle conditions, explaining how Li-Fraumeni sufferers and their p53-null mouse counterparts can function perfectly well until they encounter situations in which their DNA can be damaged. In spontaneously arising tumours, p53 activity is sometimes lost, just as in Li-Fraumeni syndrome, but it is also frequently mutated into the oncogenic versions that so confused the early researchers in the field. Mutant oncogenic p53 acts as a dominant-negative protein, locking onto any remaining normal p53 and preventing it from working.

Understandably, the p53 pathway has become a prime target for anticancer therapeutics. Although much effort has been expended in trying to design drugs to restore p53 to normality, transcription factors are notoriously hard drug targets, and progress has been slow. However, the prize is so large that the search continues, and it is likely that one or more of the increasingly ingenious strategies will eventually succeed.

And what of David and Lionel? Unsurprisingly, David never went back to immunology and has carried on working on p53 in a career that has turned him into one of the world's most prominent cancer biologists, and the second most highly cited medical scientist in the United Kingdom in the last decade. He has had an amazing number of very high profile jobs, won a ton of scientific awards, was knighted in 2000, and has been elected to every British society honouring the great and good of science. The bling, however, is not what continues to drive him: "Science is not really a career but much more an obsession—it's the science that matters above everything. The papers, the grants, the jobs, the prizes are all just surrogates for the science. At times I have been scooped or not followed a line of research that I should have, but in the end, if you honestly do your best, are very persistent, try hard to make a contribution and are generous to others, you will likely succeed" (Lane 2006).

Lionel has had a lower profile. In 1988, he returned to his Cambridge roots, running an ICRF-funded tumour virus lab in the Pathology Department until his retirement in 1995. His subsequent research moved away from SV40 to human papillomaviruses, switching from trying to understand cancer to building on current knowledge to prevent it. It was a worthwhile change, because Lionel's Cambridge lab was the intellectual birthplace of Human Papilloma Virus vaccines; one of his postdocs, the late Jian Zhou, together with a sabbatical visitor, Ian Frazer, began the work that was the breakthrough for their subsequent development of the Gardasil anti-HPV vaccine, since administered to millions of girls and women to protect them from cervical cancer. Lionel, like David, is a fellow of multiple learned societies, and in 2005, in recognition of his

Lionel Crawford in retirement.
(*Photograph courtesy of Lionel Crawford.*)

lifetime's work on DNA tumour viruses, he was awarded the Gabor Medal of the Royal Society. Described by one of his former colleagues as one of the unsung heroes of British biochemistry and molecular biology, it is only appropriate that the last word in this chapter goes to him: "The way science grows is in the cracks between the paving stones. I was a biochemist and David was an immunologist and we thought differently, sometimes at cross-purposes but sometimes synergistically. It was an awfully long time ago, but sometimes it still seems very close."

Further Reading

Lane D, Levine A. 2010. p53 Research: The past thirty years and the next thirty years. *Cold Spring Harb Perspect Biol* **2:** a000893.

Soussi T. 2010. The history of p53. A perfect example of the drawbacks of scientific paradigms. *EMBO Rep* **11:** 822–826.

Two accounts of their research from the horses' mouths.

Quotation Sources

Mitchison, Avrion. Web of Stories interview. http://www.webofstories.com/play/52547?o=S&srId=397632.

Lane DP, Robbins AK. 1978. An immunochemical investigation of SV40 T antigens. 1. Production properties and specificity of rabbit antibody to purified simian virus 40 large-T antigen. *Virology* **87:** 182–193.

Lane DP, Crawford LV. 1979. T antigen is bound to a host protein in SV40-transformed cells. *Nature* **278:** 261–263.

Lane DP. 1992. p53, guardian of the genome. *Nature* **358:** 15–16.

Lane DP. 2006. Such an obsession. *Cancer Biol Ther* **5:** 120–123.

CHAPTER 4

Country Life
Repair and Replication

For the last quarter of a century, the Clare Hall Laboratories have been a highly productive focal point for research into DNA repair and replication. Never numbering more than 10 small groups, scientists at Clare Hall look out over a landscape of fields, farms, and riding stables, shortly after the last suburbs of North London give way to the placid Hertfordshire countryside. The muddy tranquillity of their surroundings is only slightly disturbed by the distant hum of traffic from the M25 and the A1 motorways, which pass nearby. It is a far cry from the urban buzz surrounding Clare Hall's metropolitan big brother in Lincoln's Inn Fields.

By rights, Clare Hall should not have been a success. It was isolated, too small, too focussed—any of these factors could have been a death sentence. Although the importance of the steady financial support of the ICRF and, latterly, Cancer Research UK, should not be underestimated, the many achievements of those working there are a testament to motivated, smart science, fierce collegial loyalty, and a dedicated and exacting Director.

Unfortunately, a chapter extolling the talents and virtues of all of the inhabitants of Clare Hall would be impossibly long and most likely very tedious; a litany of success can be as good a soporific as any other list. The four protagonists in the pages that follow have therefore been selected for two reasons. Firstly, the two men responsible for the birth and development of Clare Hall were both in the vanguard of the DNA revolution, and their work marches with the evolution of research into replication and repair. Secondly, the stories that complete the chapter are perfect illustrations of what Clare Hall does best: hiring very good people, and then trusting their abilities to turn ambitious, long-term projects from ugly ducklings into swans.

John Cairns and Mill Hill, 'That Awful Place'

Clare Hall rose from the ashes of the ICRF Mill Hill Laboratory, a suburban outpost situated in North London at the far end of the Northern Line. Hunkered down on the doorstep of the giant MRC National Institute for Medical Research, the Mill Hill laboratories had been rented from the MRC on a 50-year lease in 1936, and for a while had been quite successful. However, like the rest of the ICRF, by the time of Michael Stoker's appointment as Director in 1969, the labs had fallen into a sleepy complacency, enlivened only by the daily ritual of Afternoon Tea in the library, complete with crustless cucumber sandwiches, and the arrival and departure of the unit Director in his Bristol 405, a sleekly beautiful four-door sports saloon that was its owner's pride and joy. The lab was rudely awakened from its genteel slumbering in 1973 by the arrival of a molecular biologist eager to turn his considerable talents to cancer research. John Cairns, the ex-Director of the Cold Spring Harbor lab, was about to shake things up rather spectacularly, in the process laying down the foundations on which Clare Hall's future successes were built.

John Cairns is a fascinating figure in the landscape of 20th century biology. Switching fields with the ease of a small boy scrambling over a fence, he left his mark on everything to which he turned his hand, leaving a trail of significant papers in each discipline before moving on to the next. As an added bonus, he is one of the most thoughtful and elegant chroniclers of the molecular biology revolution, in which he was both a distinguished participant and an observer, and his subsequent writings on the biology and epidemiology of cancer are still relevant and thought-provoking today. The esteem and affection in which he and his wife Elfie are held by their many friends and acquaintances are a tribute to their unique combination of charm, intelligence, and hospitality.

John Cairns, 2002.
(Photograph courtesy of CRUK London Research Institute Archives.)

The grandson of a Master of Balliol, son of the hugely talented and influential brain surgeon Sir Hugh Cairns, godson of Sir Charles Sherrington (Nobel Laureate and founder of neurobiology), John was born into the intellectual aristocracy of Britain and had the aptitude and connections to have followed his father into a stellar career as a surgeon. Instead, after qualifying as a doctor, he turned his back on silver spoonery, and by the mid-1950s was a successful academic virologist at the newly established University of Canberra.

In 1957, John left Elfie and his young family in Canberra for a four-month sabbatical at Caltech, and by a combination of good luck and poverty, found himself at the epicentre of the DNA revolution. Unable to afford the exorbitant room rates at the Caltech Faculty Club, John was rescued from homelessness by Jan Drake, one of Renato Dulbecco's PhD students, and as a result found himself eating dinner most nights in an apartment shared by Drake, Matt Meselson, and Howard Temin during the period in which Meselson and Frank Stahl were doing what Cairns subsequently christened "the most beautiful experiment in biology."

Meselson and Stahl's 1958 paper, "The Replication of DNA in *Escherichia coli*," was the first big step towards the wider acceptance that DNA really was the hereditary material of the cell, and not just a boring structural component of chromosomes, as many of the biochemists and physical chemists who worked on it had previously decided. Oswald Avery, Colin MacLeod, and Maclyn McCarty's 1944 experiment showing that the "transforming principle" of *Streptococcus pneumoniae* was DNA, and Al Hershey and Martha Chase's more recent demonstration in 1952 that the infectious material of a phage was also DNA, had not made much of an impact outside the world of molecular biology, perhaps being regarded with suspicion as new-fangled genetic sleights of hand. Watson and Crick had suggested in 1953 that the invariant base pairing of the rungs of the DNA ladder, with cytosine (C) always with guanine (G), and adenine (A) always with thymine (T), provided a great mechanism by which one strand could produce a complementary copy of itself, but it was just another theory until Meselson and Stahl gave it reality. (It should be noted that the invariant C:G and A:T ratio between the bases was originally observed by Erwin Chargaff, but he failed to realise the significance. He was, consequently, very grumpy about DNA and Watson and Crick for much of the rest of his life.) Although they were careful not to overinterpret their experiment, referring cautiously to two DNA *subunits*, Meselson and Stahl's work showed that the two strands of the DNA double helix came apart and were each used as a template to make a new strand. The result was a duplication of the original molecule, with each duplicate containing one old and one newly synthesised strand; in other words, DNA was replicated semi-conservatively.

John Cairns had a wonderful time at Caltech. He got to grips with molecular biology whilst doing a lot of washing up at the Meselson ménage, attended the phage course (he is remembered by one of the other participants as "admirably modest, generous and collegial, but [...] it was rather as if Sir Lancelot had decided to attend the school prom") (Susman 1995), and survived an attempt at manslaughter by camping, Delbrück style ("as I waited through the night for the dawn to come, I thought I would freeze to death") (Cairns et al. 2007). It was not entirely surprising that upon his return to Canberra, he was an addicted molecular biologist and switched the focus of his science accordingly. Following another, longer sabbatical with Al Hershey at Cold Spring Harbor in 1960, John published the definitive proof that DNA replicated as a single double helix and that replication in bacteria proceeded bidirectionally round the circular *E. coli* chromosome from a single origin. In the process, he invented the technique of DNA autoradiography, visualising DNA by radioactively labeling it with tritiated thymidine. His autoradiographs of *E. coli* DNA caught in the act of replication are simultaneously a technical tour de force and a monument to extreme patience and persistence. John himself dismissed this minor classic as "an act of tidying (rather like washing-up, in fact)" (Cairns 1980).

In 1963, John moved his family permanently to Cold Spring Harbor because he'd been inveigled into becoming the laboratory's Director, little realising that the place was on its knees both financially and structurally and that he had been hired to oversee its downfall. As Matt Meselson recalls: "I think John was expecting the same gentlemanly standards that John himself would have practised; instead he walked into a nightmare. But he did get the place out of the red, which was an enormous accomplishment. It was about to expire. Not only expire, but to expire in sewage" (Meselson, CSH Oral History). After five years of "the most hideous time," during which he greatly annoyed the laboratory's detractors by getting the place back on its feet again, John handed over the Directorship to Jim Watson and got back to what he loved best, doing experiments.

The project that John began impinged on a parallel, biochemical strand in the story of DNA—the question of how the molecule was made. Although Meselson and Stahl had shown that semi-conservative replication could occur, the nuts and bolts of exactly how DNA replicated itself was a very biochemical matter. Where there is a process, there has to be an enzyme to perform it, and fittingly, the eureka enzyme of DNA replication, DNA polymerase, was discovered by one of the greatest biochemists of the 20th century, Arthur Kornberg.

Kornberg, as the host of stories lovingly propagated by his friends and colleagues testifies, was in his own way as charismatic as Delbrück but was an implacable dragon in the laboratory. His allegiance to biochemistry meant that anything other than total commitment and experimental rigour was unacceptable, encapsulated in the maxim that he adopted from his fellow biochemist Efraim Racker: "Don't waste clean thinking on dirty enzymes"; and the first of his famous Ten Commandments of Enzymology: "Thou shalt rely on enzymology to resolve and reconstitute biologic events." Kornberg's genius was to realise that everything, even the most complex molecular machine, could be purified out of the cell, and that reconstituting such machines in the relative tranquillity of the test tube was the way to understand how they worked. His interest in DNA arose because he wanted to know how nucleotides, which he and others had shown were the building blocks of DNA chains, were put together. The fact that DNA might be the hereditary material and that Watson and Crick had proposed a model for its double-helical structure was interesting, but the possibility of proving that DNA synthesis was not one of life's untouchable mysteries, but as much an enzymatic process as the fermentation of yeast, was what initially drove him. From 1954, when he was appointed Chair of the Microbiology Department at Washington University in St Louis, Kornberg worked towards detecting the polymerisation of nucleotides, beginning first with RNA and then widening his studies to include DNA synthesis.

After some years of labour, Kornberg and his lab got a hint that they were on the right track when they managed to get an *E. coli* cellular extract to synthesise a vanishingly small amount of polymerised DNA. This was all Kornberg needed. Two years later, when he won the 1959 Nobel Prize along with his mentor Severo Ochoa, "for their discovery of the mechanisms in the biological synthesis of ribonucleic acid and deoxyribonucleic acid," he recalled that "through this tiny crack we tried to drive a wedge, and the hammer was enzyme purification" (Kornberg 1960). The enzyme his lab purified, DNA polymerase, was able to stitch together nucleotides to make a new DNA strand, but crucially, only in the presence of an existing strand; the substrate, DNA, was directing its own synthesis, an event unprecedented in the annals of enzymology, but exactly as the semi-conservative replication observed by Meselson and Stahl predicted.

Up until John Cairns's release from the burden of the Cold Spring Harbor Directorship, everybody believed that the Kornberg DNA polymerase was the solo star of the replication show, doing everything by itself. John and his technician Paula deLucia changed all this by showing that although the Kornberg DNA polymerase might be able to synthesise DNA in a test tube, it was certainly not doing the real work of replication in *E. coli.* In so doing, they scored a satisfying home run for molecular biology, showing to the biochemists that bacterial genetics was as important a tool as enzyme purification.

Having heard that Roy Curtis at Oak Ridge National Laboratory had isolated a mutant of *E. coli* that made miniature daughter cells containing no DNA but lots of Kornberg polymerase, John set Paula the difficult task of finding the reciprocal mutant of *E. coli,* one that lacked Kornberg's polymerase but was perfectly viable. This she did, and the mutant, which John christened polA in homonymic tribute to Paula's technical virtuosity, caused a flurry of biochemical activity resulting in the isolation of two further *E. coli* DNA polymerases, one of which, DNA polymerase III, was the real replicative enzyme. Arthur Kornberg's chagrin that a mere molecular biologist had managed to show something so important without so much as partially purifying a protein, was no doubt mollified by the fact that both new polymerases were identified by his son Tom, whilst a graduate student in Malcolm Gefter's lab at Columbia.

The business of bacterial DNA replication was not to occupy John for much longer. Keeping Cold Spring Harbor afloat had comprehensively emptied the coffers of the Cairns family, and in 1972, John started job-hunting back in the United Kingdom, where his children could finish their educations for rather less than in the States. In order to be on the spot, he arranged a year's sabbatical with Michael Stoker at Lincoln's Inn Fields. It was not long before Stoker realised that John was the perfect solution to a problem that had been dogging him for some time—whom to appoint as the new head of "that awful place" Mill Hill.

In John's words, Mill Hill "was full of people who had decided that safety lay in not doing anything [...]. I had to terrorise them, but after Cold Spring Harbor that was easy peasy!" The new laboratory Director, in addition to frightening the life out of the old staff, abolished the cucumber sandwiches and the segregated tea rooms and imported a whole slew of young, ambitious developmental biologists, whose story is related in detail in **Chapter 9**. John also changed the direction of his own research once again and started thinking in depth regarding cancer.

The switch in emphasis gave rise to a rather good book, *Cancer: Science and Society*, published in 1978, and yet another distinguished career in cancer epidemiology and public health research. In the short term, however, John's work on DNA replication in bacteria morphed into studying another aspect of the molecule's lifestyle: how it repaired itself when damaged.

The mere existence of DNA repair had come as a big surprise to many people. In the first half of the 20th century, genes were a concept, rather than a fact, and working in such an information vacuum meant that some interesting misconceptions had arisen. One such, in part propagated by the 1935 paper by Max Delbrück, Nikolai Timoféeff-Ressovsky, and Karl Zimmer that had inspired Schrödinger's *What Is Life?*, was that it was incredibly hard to mutate genes. To create any kind of change, exposure to extreme stress such as high-intensity ionising radiation was thought to be necessary, because genes were viewed as extremely stable entities, heavily protected from the outside world in order to keep their precious cargo of genetic information intact. This was backed up by the very low rate of spontaneous mutation seen in the organisms, such as the fruit fly, in which the early geneticists did most of their work.

Delbrück and his colleagues had been heavily influenced by discussions with Hermann Muller, who in 1927 had been the first to show that X rays produced high mutation rates in fruit flies. Mutation being the lifeblood of genetics, Muller's findings had been taken up with great enthusiasm by his field as an extraordinarily powerful tool (he was eventually awarded the 1946 Nobel Prize for Physiology or Medicine). As a spin-off from Muller's work, the study of how radiation affected living things had become a discipline in itself by the 1930s, and the field expanded hugely over the following decade, thanks to the U.S. Atomic Energy Commission, which lavishly funded radiation biology during and after the Second World War in an effort to find ways to counteract the lethal effects of the atomic bomb. However, there were problems with radiobiology. Although it was plain that ionizing radiation, and also ultraviolet light and some particularly nasty chemicals such as mustard gas, must be damaging genes, nobody had a clue how. Matters were not helped by the field itself. Innovative and thoughtful radiation biologists did exist, but the excessively generous funding had resulted in an accumulation of mediocrity, with some labs content to repeat the same experiments again and again for no reason other than that they counted as data, however useless. Radiobiology gained a reputation for existing in a state of bloated torpor, and research into the nature of genetic mutation slumbered alongside it.

What everybody had missed, and could probably not have predicted before DNA came onto the scene, was that far from being swaddled tenderly in the warm, cosy environs of the nucleus, DNA was continually buffeted by storms of passing mutagens. The reason for the low rate of spontaneous mutation was not that genes existed behind a firewall, but that they were dynamically stable. Like a circus plate-spinner, the cell keeps its genetic information intact because it continually surveys its DNA, detects any damage, and repairs it. Jumping ahead in time, David Lane and Lionel Crawford's p53 protein is a sophisticated gatekeeper of this determination to preserve an unblemished genome.

Because sunlight and life go hand-in-hand, enzymatic photoreactivation of DNA after ultraviolet (UV) irradiation is probably the oldest repair mechanism in existence. It was also the first identified, stumbled on inadvertently by Albert Kelner at Cold Spring Harbor, and Renato Dulbecco during his time in Salvador Luria's lab in Bloomington, Indiana. Both Kelner and Dulbecco had been greatly frustrated by their attempts to examine the survival of, respectively, bacteria and phage, following UV irradiation. Both men found that their experiments were completely irreproducible. Sometimes, irradiation resulted in perfect survival curves, where colony number was inversely correlated with the intensity of UV treatment, but sometimes, the curves were all over the place, with seemingly random numbers of survivors cropping up even after punitive UV treatment. After months of checking and rechecking his experimental protocol, Kelner in desperation moved his bacterial plates into a cold room, so that he could work out, using a series of water baths, whether temperature played any part in the process. To his surprise, all of the survival curves began to behave perfectly, but shortly afterwards, the penny dropped: The cold room was windowless, in contrast to his previous lab, and the more sunlight the plates saw, the more colonies grew on them. Dulbecco, with the same problem in his phage experiments, and knowing of Kelner's work, came to the same conclusion, that daylight could induce repair of UV damage, and their two papers on the subject were published in 1949.

That DNA was the substrate for photoreactivation was not revealed until 1956, by Sol Goodgal and Stan Rupert at Johns Hopkins in Baltimore. The two men showed that extracts of *E. coli* cells, when mashed up and added to UV-irradiated DNA from the bacterium *Haemophilus influenzae*, were able to repair the UV damage, as long as the reaction was exposed to daylight. The inducible repair activity in the *E. coli* extract was clearly due to an enzyme of some kind, because it could be killed by heating, but it took until 1978 to find what the enzyme was, as it was present in such small quantities that it proved almost impossible to purify. In the meantime, however, biophysical analysis had shown that the lesion in DNA on which the mystery enzyme acted was a pyrimidine dimer, an unholy alliance formed between two neighbouring cytosine or thymine bases on the same strand of DNA. Such dimers were very common and were dangerous because they were unable to base-pair properly, leading to errors of replication, and hence mutation. Photoreactivation repaired pyrimidine dimers by resetting them back into their monomeric states, where they were once more able to recognise their partners on the opposite strand of the double helix and could be replicated correctly.

Photoreactivation was just the tip of the repair iceberg. In 1964, Dick Boyce and Paul Howard-Flanders at Yale, and Bill Carrier and Dick Setlow at Oak Ridge National Laboratory, with several other labs hot on their heels, independently discovered an alternative mechanism for ridding the cell of pyrimidine dimers: nucleotide excision repair. Rather than rescuing the bases by remonomerising them, cells could simply cut them out, together with a short stretch of the sequence in which they were embedded, leaving a gap that could be filled in using the opposite strand of the double helix as a template.

Excision repair turned out to be able to fix a lot of other lesions besides pyrimidine dimers, and its discovery prompted a new theory to explain why DNA was double-helical. As those working on the biochemistry of replication and transcription of DNA were discovering, unwinding the strands of a double helix is a difficult topological feat; simply loosening the twist in one place would result in tightening the twist in another, and dissecting the enzymatic contortions that cells have evolved to avoid hopeless tangling has kept a legion of biochemists occupied for many years. Why would a cell go to the bother of a double helix at all, when one copy of the genetic information per chromosome was surely enough? The repair field thought it had the answer. In a *Scientific American* article from 1967, Bob Haynes and Phil Hanawalt suggested that "redundancy is a familiar strategem to designers of error-detecting and error-correcting codes. If a portion of one strand of the DNA helix were damaged, the information in that portion could be retrieved from the complementary strand" (Hanawalt and Haynes 1967).

Asserting that the mighty double helix existed because it was the best way to correct errors in the genome was testimony to the excitement in the new field. Shedding many of the outdated concepts of radiobiology, repair enthusiasts had realised that the discovery of DNA-specific enzymes, such as Kornberg's DNA polymerase and the DNA-cleaving nucleases, meant that the genome was as subject to enzymatic modification as any other cellular substrate, and therefore many types of enzymatically driven repair mechanisms were likely to exist. Those prepared to branch out into a bit of genetics, to find useful mutants, and biochemistry, to characterise the processes they revealed, began mining a rich seam of unknown enzymes and pathways.

In 1967, DNA repair made it into humans, when Jim Cleaver, working at the University of California, San Francisco, realised that xeroderma pigmentosum (XP), a genetic disease causing sun sensitivity and predisposition to skin cancer, was caused by defects in excision repair. The discovery provoked elation in everybody working on DNA in bacteria, from the molecular biologists right through to the biochemists, whose sentiments were encapsulated by Bob Haynes: "I can vividly remember saying to several people, thank God, we've now got a disease. We'll be able to get money from the NIH because DNA repair is relevant to a human disease!" (Friedberg 1997).

Rather surprisingly, the link between DNA repair and cancer was not pursued with much enthusiasm until nearly a quarter of a century later, when, as we shall see below in this chapter, DNA repair defects were shown to be major contributors to colon and breast cancer susceptibility. Although there were some technical obstacles, the main reason appears to have been a lack of communication between fields. Theodore Boveri had proposed that cancer was a disease caused by mutation of the genome back in 1914, but the molecular biologists were preoccupied with understanding cancer through the oncogenic tumour viruses, and even if they knew about it, were not particularly interested in Boveri's elderly somatic mutation theory. Peter Brookes and Phil Lawley had shown that DNA was the target of carcinogens in the early 1960s, but they were medicinal chemists, and their field was oriented towards finding and testing new carcinogens and anti-carcinogens, rather than wondering about the biology of cancer. Furthermore, they were in London, and their work was largely overlooked in America. Finally, the traditional radiobiologists were too old-fashioned to be relevant, and the few that were any good were mostly working on the mechanisms of DNA repair in bacteria and lower eukaryotes, where both biochemistry and genetics could be deployed.

There were a few lone voices. Back at Mill Hill, John Cairns had not joined the molecular biology stampede into tumour virology but was bucking the trend to think instead about the unfashionable topic of somatic mutation and cancer. In 1973, Bruce Ames had devised his eponymous test, in which potential carcinogens were assayed for mutagenicity in bacteria rather than in animals. Ames's work brought into focus the long-standing observation that mutagenesis and carcinogenesis are not always linked; the incidence of cancer in a particular species is not solely determined by exposure to mutagens, and the most powerful carcinogens are not always powerful mutagens. Clearly, something else was going on. In addition to the obvious, that there might be a disparity in DNA repair efficiency, John came up with the Immortal Strand Hypothesis, the idea that all stem cells carry an original, error-free copy of the body's DNA and that they achieve this by asymmetric replication, passing on newly synthesised strands, into which the error-prone polymerases might have inserted mistakes, to their less important progeny. The theory, although probably wrong, greatly stimulated thinking about how stem cells might retain their identity, and it was in its pursuit that John and his PhD student Leona Samson fell into the DNA repair field, where they were to discover an entirely new enzymatic process.

Using a method originally designed to determine whether bacteria replicated asymmetrically (they didn't), Leona started testing whether the mutation rate of bacteria was proportional to the mutagen concentration, using an alkylating agent, *N*-methyl-*N*′-nitro-*N*-nitrosoguanidine (MNNG), which works by sticking methyl groups onto DNA. Depending on where the methyl group is put, the effect of MNNG ranges from mild to severe; in the worst-case scenario, a guanine residue can be turned into O^6-methylguanine, which a DNA polymerase can mistake for adenine, meaning that a thymine instead of a cytosine residue is incorporated into the opposite strand of DNA during replication.

Leona was convinced for some months that she was messing up her experiments, because addition of MNNG to the bacteria had a very unexpected result. Although mutation rates in the first hour were, indeed, proportional to the concentration of MNNG, after this time the bacteria stopped mutating and acquired some kind of immunity to MNNG. Eventually, having discounted any possibility of incompetence, Leona and John realised that the bacteria must be switching on a DNA repair mechanism in response to addition of the carcinogen. The adaptive response to alkylation damage was a completely new phenomenon, and following Leona and John's 1977 *Nature* paper describing it, John, Leona, and their colleagues Penny Jeggo, Martine Defais, Paul Schendel, and Pete Robins went on to show very elegantly that the repair reaction was extremely odd. Firstly, the reaction worked really well up to a certain concentration of MNNG, but there was an upper limit above which the bacteria surrendered and started mutating again, suggesting that whatever was responsible was present in limiting amounts in the cells. Secondly, O^6-methylguanine vanished into thin air with astonishing rapidity when adapted bacteria were challenged with MNNG. In 1979, explanations for these two oddities were proposed in two back-to-back *Nature* papers. John, together with Pete Robins, proposed that the kinetics of repair that they had observed were due to the responsible enzyme only being able to act once, and a complementary paper, from the Swedish biochemist Tomas Lindahl's lab, showed that the O^6-methylguanine simply reverted back to guanine in a single step. Whatever the repair mechanism was, it was going to be a novelty.

By the time the Robins and Cairns paper was published, some unsettling nonscientific events were also occupying John's time. By the end of the 1970s, it was clear that there was no future for the ICRF in Mill Hill, because the MRC were determined to take the building back upon the lease's expiry in 1986. Walter Bodmer, who took over from Michael Stoker as ICRF Director in 1979, remembers that when he "talked to the MRC about the end of the lease, they insisted on having it back as agreed, and so got a bargain" (the original agreement stipulated that the MRC would pay the ICRF a mere £10,000 upon the reversion of the lease). The unit's scientists were understandably greatly dismayed. John Cairns had succeeded in a very short time in rekindling the Mill Hill laboratory's reputation as a place where good science was done, and his developmental biology initiative was blossoming spectacularly, so everyone there was determined to try to stick together, come what may. Unfortunately, the price for this decision was the loss of the unit director. John recalls:

> I had to worry about what the hell was going to happen to all the very bright young scientists who were working in the building. So we had a Quinquennial Review [the five yearly transit of Hell by which scientists decide whether their peers are any good or not] with Matt Meselson and Sydney Brenner and a few other people, and they decided that the

> solution to this problem was that I should resign from being Director so that a younger Director could be brought in. This would force the ICRF to fund some kind of continuation for the whole unit. So my retirement was simply to benefit the people I had brought in (Cairns, CSH Oral History).

John left for the Harvard School of Public Health in 1980, in retrospect a bad idea, because he has described it as "something of a hell hole." He could have stayed in England: "the other option was to go to Cambridge and be with Sydney [Brenner]. And I had a date to go and talk to Sydney about this, and I just damn well forgot!" British science lost the opportunity for a memorable double act.

John's strategy of falling on his sword may not have provided an optimal outcome for him, but it had the desired wider effect, saving the jobs of his younger colleagues and paving the way for his successor. Very luckily for the ICRF, romantic necessity had already managed to recruit an eminently suitable candidate as John's replacement: Tomas Lindahl, the biochemist who was in the process of nailing how John's Ada protein worked, had fallen in love with Beverly Griffin, still working on the Fifth Floor at Lincoln's Inn Fields, and was looking to move from Gothenburg in Sweden to join her in London.

Phoenix Rising—Tomas Lindahl and the Birth of Clare Hall

In what seems to be a recurring theme for repair aficionados, Tomas Lindahl stumbled into the field that became his life's work whilst trying to study something else. Stockholm born and bred, Tomas trained as a medic but was lured into research by Einar Hammarsten, Emeritus Professor of Biochemistry at the Karolinska Institute and an influential pioneer in nucleic acid research. In the mid-1960s, Tomas left Sweden for a postdoc with Jacques Fresco at Princeton, to work on heat-induced unfolding of transfer RNA (tRNAs are the group of molecules that bring amino acids to the ribosome factories where new proteins are built). In contrast with DNA, RNA work is a nightmare for the sloppy scientist, because RNA is extremely prone to being eaten by the ravening hordes of ribonucleases that seem to hover around every laboratory bench. Therefore, the Fresco lab regarded the new postdoc's failure to get his experiments to work as a sure sign of incompetence, probably brought on by the double handicap of being both Swedish and an MD. The new postdoc, however, had other ideas: "I had made that RNA … and I knew it wasn't contaminated.

I did the experiments over and made the trivial observation that the tRNA always decomposed at the same rate. If it was just due to an accidental nuclease contamination then different preps would be different, but it always decomposed at the same rate. I even wrote a little paper on that which was so boring my boss didn't put his name on it."

Tomas's discovery, that the tRNA, rather than being chewed up by ribonucleases, was spontaneously decomposing as a result of being heated during the experimental protocol, prompted him to wonder whether DNA might also be prone to such damage:

> I thought that if a small molecule like RNA actually decomposes at a rate that you can see in the lab, what about DNA? It's supposed to be a bit more stable but still, perhaps DNA is less stable than people have been thinking about. Nobody was working on that so I put that aside, and just to show the slow pace of research in those days, I went to Rockefeller and was a postdoc there for two years, and then I went back to Sweden and set up my own lab there and started doing some enzymology on enzymes like DNA ligases. Three years after I'd done those initial experiments in the US, I thought: "I'll go back and have a look at the stability of DNA because nobody else is doing it."

There is a special pleasure in a good idea simply lying around waiting for an astute person to pick it up, especially when it goes against the tide of general opinion. Tomas's early realisation, and subsequent career-defining demonstration, that DNA was inherently unstable and could be damaged without any requirement for exogenous mutagens, solved a problem that should have been perplexing the DNA repair field, but that they had unaccountably missed: what was the original purpose of the clearly ancient repair mechanisms? Photoreactivation was explicable, because the threat, sunlight, remained the same, but the other types of repair were different. Cells might be able to fix the DNA damage perpetrated by intense ionising radiation or doses of synthetic chemical carcinogens, but these very modern insults could not have driven the evolution of the repair machinery; a much older way of damaging DNA had to be responsible. Tomas, with his solid grounding in nucleic acid biochemistry, was the first to understand that the threat lay in the composition of the DNA molecule itself; by their very nature, some of the chemical bonds formed inside the bases, and between the bases and the sugar–phosphate backbone, are relatively susceptible to breakage. Cytosine, for example, can be readily deaminated, that is, stripped of its amino group, which turns it into the simpler base uracil. Uracil can pair comfortably with guanine, leading to the same errors in replication and transcription induced by chemically and radiation-induced damage to cytosine bases.

In tracking down how the cytosine-to-uracil error was fixed, Tomas and his lab unmasked an entirely new class of repair mechanism: base excision repair. Uracil is detected and chopped out of the DNA chain by a glycosylase, an enzyme able to cleave the bond between the errant base and the backbone, leaving the double-stranded backbone intact. A gang of other enzymes then takes over, chopping out the sad remains of the original nucleotide together with its nearest neighbours, mending the gap using the opposite strand as a template, and finally stitching up the break.

Tomas's strategy of looking at repair as a response to intracellular rather than extracellular insults was an innovative approach that threw up numerous other ways in which DNA can be damaged just by the cell going about its normal daily life. However, these attacks by the toxic by-products of metabolism are valiantly fought off by an arsenal of DNA repair mechanisms that keeps the genome in surprisingly good shape; a recent paper comparing the genomes of parents and offspring showed that of the 3 billion base pairs of human DNA, only around 70 had been mutated between the generations.

Somewhat remarkably, whilst Tomas was making a name for himself working on the enzymology of DNA repair, he had also managed to become a big noise in the tumour virus world. As he recalls: "I was doing two careers in parallel because I was doing the work I liked, DNA repair and DNA metabolism, and my chairman in Stockholm, Peter Reichard, came and said 'That's all very well, but nobody else in Sweden is doing that, so it would be good if you did something else so you can talk with people.' " Reichard suggested a collaboration with the eminent virologist Georg Klein, a fellow Karolinska inmate, who worked on Epstein-Barr Virus (EBV), and thought the time was ripe for someone to start some EBV molecular biology. The collaboration quickly bore fruit, with Tomas making the major discovery that EBV replicated in cells as a discrete circle, rather than by the SV40 and polyoma strategy of integrating into host DNA.

In 1978, Tomas left his friends and family behind to move across the country from Stockholm to Gothenburg, where he had been offered a job as Professor of Medicinal Chemistry. However, although he had a large U.S. grant and a small teaching load, and both branches of his scientific life were going swimmingly, he only remained in Gothenburg for three

Beverly Griffin and Tomas Lindahl, 2002.
(Photograph courtesy of CRUK London Research Institute Archives.)

years. In that British bastion of tumour virology, the ICRF, Beverly Griffin had also started working on EBV, and a cross-border romance had sprung up whose importance far outweighed any academic considerations.

Coming to London to be with Beverly turned out to be a good career decision as well as the start of a life partnership that has endured to this day. Because the ICRF Director Walter Bodmer was a human geneticist, he was more clued up than most cancer biologists about the importance of somatic mutation and therefore DNA repair, and willingly subscribed to the idea that it was a good area for a cancer charity to fund. After a short time on the Fifth Floor with Beverly, Tomas set up a lab at Mill Hill, taking over John Cairns's space there. He also reverted to monotheism, stopping his EBV projects to work entirely on DNA repair: "For a happy year or two, we discovered about one important DNA repair enzyme every year, and I thought: 'This won't last forever, this actually creates a new branch of the DNA repair field that will hopefully become important.' So I decided rather than getting an ulcer and trying to be on top of two fields at the same time, to put viral work aside."

One of the first fruits of this decision was a paper describing the enzyme responsible for the adaptive response to alkylation first observed by Samson and Cairns. Tomas's Mill Hill group showed in 1982 that the enzyme, O^6-methylguanine-DNA methyltransferase, restored mutant O^6-methylguanine bases to health by removing the methyl groups and disposing of them by methylating itself, the enzymatic equivalent of falling on a live grenade. Suicide inactivation, as the mechanism was christened, was not the only trick this enzyme had up its sleeve; subsequent work from Tomas's lab showed that methylation at a different position in the protein turned it into a transcription factor that could not only activate its own gene, but also several others in the adaptive response pathway. Tomas is very proud of this work, published in 1986 in *Cell*: "If I may say so, I think it's a well written paper and it had some impact! If I was going to mention one paper which I wrote myself ... which I actually feel satisfied with, where I actually got things right, and I didn't say too much or too little, it's this one."

Tomas clearly being a good researcher, Walter was very anxious to keep his new recruit happy and to involve him in the ongoing problem of what was going to happen to the scientists at Mill Hill. In 1982, Walter cut a deal that allowed the Mill Hill developmental biologists to move into a new, ICRF-funded unit at Oxford University, but the problem remained of what to do with Tomas. Shoehorning him into Lincoln's Inn Fields was not an option that Walter, always an enthusiastic expansionist, favoured, and the healthy economic climate meant that there was money to spare for something more ambitious. The animal and service facilities at Mill Hill and Lincoln's Inn Fields were going to be unified and expanded on the site of Clare Hall, an empty and decaying ex-smallpox and TB hospital near the newly built M25 motorway, so perhaps it would be possible to persuade Tomas to go there too?

Unlike John Cairns, who had thought that Clare Hall was far too isolated to work as a research laboratory, Tomas could see that there were possibilities out in the country, as long as several conditions were met. The first was that the laboratories should be big enough to accommodate a critical mass of scientists, which meant expanding the existing building plans, something Walter was happy to do. Owing to planning restrictions imposed by the local council, the laboratory could only be two storeys high at most, but this suited Tomas, who had observed in Sweden that "for some crazy reason scientists communicate very well horizontally and very poorly vertically. It sounds silly but for some reason a set of stairs is a really effective barrier, and walking down the corridor is much easier. And you can see that at Lincoln's Inn Fields—the fifth floor and the fourth floor and the sixth floor were just different empires and didn't actually chat. People on the same floor interacted closely and aggressively but at least interacted!" Accordingly, the Clare Hall labora-

Clare Hall laboratories.
(Photograph courtesy of CRUK London Research Institute Archives.)

tories took shape as an elegant linked collection of five low buildings, three devoted to research, and two to services, surrounded by landscaped gardens, all fitting rather well into the rural setting.

Having achieved the building he wanted, Tomas, now officially appointed Director of the Clare Hall Laboratories, decided that the labs would be very focused: "I was very concerned it wouldn't be like the classical biochemistry or cell biology department that you see in universities, where because of the teaching, you have to be very broad like an old fashioned zoo—one zebra, one lion, one giraffe and one turtle and so on! That's obviously an exaggeration, but if you have as your next door neighbour an expert on lipid metabolism it's not perhaps somebody you easily share a seminar series with if you do nucleic acid or molecular biology work." Not surprisingly, the focus that Tomas chose was his own—the biochemistry and metabolism of DNA, and the three Rs of Repair, Recombination, and Replication.

Sustained institutional success in science is a rare and interesting beast, seemingly dependent on a single factor: appointing the right person to be in charge. This certainly seems to have been the case with Clare Hall. Matched only by Cold Spring Harbor in terms of the scientific quality and impact of its publications, with its occupants clanking around under the weight of their many awards and honours, Clare Hall's success revolved around its Director's abilities. Tomas had an extremely good eye for hiring stars in the making, but turning Clare Hall into a place where the stars wanted to stay long term was an even greater feat.

As Director, Tomas was a figurehead that the Clare Hall inmates could trust and be proud of. His own work was good enough that he commanded respect from the faculty, and in return, he expected very high standards of scientific integrity and rigour; you were unlikely to do shoddy or boring work at Clare Hall and get away with it for long. Perhaps even more importantly, Tomas was genuinely interested in what everyone was doing and possessed an unusually high degree of intellectual generosity; he always had time to talk with other people about their work, even if there was no benefit to his own. This fostered a tremendous atmosphere of communal thinking, where ideas could be freely debated. A scientific problem shared is not necessarily halved, but the combined weight of a group of highly knowledgeable smart people thinking about something is guaranteed to eliminate half-baked ideas and unthinking dogma, clearing the way for original and exciting experiments. And people could afford to be generous with their thoughts; perhaps by luck as well as judgement, Tomas managed to hire people who were scientifically close enough to interact fruitfully but far enough apart that they were never direct competitors, which would have irretrievably poisoned the atmosphere.

Tim Hunt, who arrived at Clare Hall in 1990, eleven years before he won the Nobel Prize in Physiology or Medicine together with Paul Nurse and Lee Hartwell (who are to be found in **Chapter 6**), is an enthusiastic fan: "Tomas was so refreshing compared with the Professors of Biochemistry I'd known; here was a real biochemist who actually cared and knew what everything was. His whole philosophy was to try to reduce the energy barrier for doing experiments. He wasn't at all manipulative or scheming. He was terribly pleased if you got a good result, although he was very hard on the people who didn't! People didn't last at Clare Hall if they couldn't cut it."

Tomas's interest in his group leaders' work extended to making sure that they were properly recognised for their achievements; thus, he was very active in nominating them for prizes and awards. He and the senior staff also made sure that junior lab heads with interesting work broke into the conference and seminar circuits as early as possible. This latter strategy was so successful that it gave rise to the mistaken impression that Clare Hall was a big institute; those not in the know assumed that the Clare Hall stars must be resting on a much larger bed of mediocrity, not realising that the stars were all there were.

A further boost to productivity was the almost complete absence of committees. Steve West, one of Tomas's earliest recruits, characterises his managerial style thus:

> Tomas was remarkable in that he never wrote an email. He would show up at the lab door and bring up something, then he'd listen to what you had to say—he'd never comment on it!—and then he'd go away and make a decision. You didn't know whether or not you influenced that decision, because sometimes it wouldn't be what you'd suggested, but he always made everybody feel that their opinion was valued, even if he completely ignored it! And that's quite a good skill. And Tomas's batting average was phenomenal in terms of making the right decisions.

Intellectual powerhouses cannot thrive long term unless their occupants are reasonably happy with their lot, and although Tomas's egalitarian management style was greatly appreciated, he had two other invaluable assets, in the persons of Brenda Marriott and Frank Fitzjohn. With Tomas, Brenda and Frank formed a triumvirate that dealt with anything getting in the way of doing good science, and their importance to Clare Hall's success cannot be overstated.

Frank's job of Clare Hall lab manager was dreamt up by Tomas after his time on the Fifth Floor at Lincoln's Inn Fields. In the frenetic atmosphere there, communal equipment was frequently broken, radioactive, or both, because nobody was prepared to take responsibility for fixing it. Anyone who has experienced the homicidal rage generated by standing in front of a broken centrifuge, holding a rack full of laboriously prepared and rapidly disintegrating samples, can understand the importance of having someone to ensure the smooth running of a lab, and Frank, in the nicest way possible, was a combined troubleshooter and vigilante without parallel. Bill House, who had hired him into the position, had simply described his job as: "Look after Lindahl," and Frank took him exactly at his word. Whatever was required, from guiding fundraisers round the labs (elderly lady: "How many people work here?" Frank: "About half of them."), to climbing into the drains to check them for radioactivity, fell within Frank's remit. The move from Mill Hill to Clare Hall was seamlessly arranged and executed by Frank (Tomas: "Move us to Clare Hall, Frank, I'll see you there in three weeks"), and thereafter, successive generations of scientists benefited from Frank's efficiency (David Lane: "The good thing about Frank was, you asked him a question and he always said yes!").

If Frank kept the place in perfect working order, Brenda Marriott, in the next office, was the guardian of its emotional well-being. Brenda's potential had been discovered early on by John Cairns, who had hired her as a junior secretary when he first came to Mill Hill. After being roundly told off by the young Brenda for writing a "beastly" letter to someone, John realised that he had acquired someone rather out of the ordinary, and promoted her. By the time he left, Brenda was running the day-to-day business of the unit, and Tomas took her to Clare Hall as his chief administrator. Although she was as efficient in this as Frank was at managing the labs, she is fondly described by Tim Hunt as being the "den-mother" of the place, the person who turned Clare Hall into a family. Brenda was the person you first met as you arrived from the airport, the one who distracted your crying children as you struggled with your bags up the stairs to the flats in Clare Hall Manor, the grand 18th-century country house next door to the labs where incoming staff were accommodated. Brenda soothed broken hearts, found dentists for broken teeth, reined back the foolhardy and encouraged the nervous, settled disputes and gave advice, and generally acted as a conduit between the lab and Tomas.

Tomas describes his two lieutenants thus:

> If Clare Hall became a success I can take some of the credit for that, but some of the credit is due to Brenda and Frank. They created [a] very efficient buffer, so when people wanted something, including lab heads, they would first go to Brenda and Frank, [who] had a very good sense when a scientist tried to take selfish advantage or when there was a real request.
>
> Brenda is a very outgoing person, and people came and told her all kinds of things. She may have told me something like 5% or 10%, and she sifted the information I needed to know, because if she had told me everything, people would quickly find out and nobody would talk with her any more! There were things that perhaps I should have known that I didn't know about, but more importantly, if there was something, she would come and say, do you know this, do you know that? And Frank was the same, and he also kept a good eye on the place, so if you found a dirty ultracentrifuge he would backtrack and find who had done that. Both he and Brenda had the sort of personalities that meant nobody on the site was keen to pick a fight with them or be told off by them. So they ran the place, and I was more the last instance when there was real disagreement.

Clare Hall wasn't perfect. Undoubtedly, if people couldn't or didn't want to fit in, they were very unhappy there and left, along with those who didn't make the grade. It wasn't ideal for PhD students, because they really did feel the isolation and missed out on the metropolitan life being enjoyed by their compatriots at Lincoln's Inn Fields. It spectacularly failed to appoint female group leaders, although Birgit Lane, one of only two women to run a lab there, charitably attributes this to benign neglect, rather than any more sinister motive. On the whole, however, it remains a shining example of how to create and maintain an environment in which intelligent, focused, important research can be performed as easily as possible.

Despite the stresses and strains of being Clare Hall Director, Tomas managed to go on working very productively. Uncovering the multitude of weird and wonderful enzymes involved in base excision repair was to take up much of his subsequent career, although he also made some notable diversions. One of these came to fruition in 1988, the year Tomas was elected to the Royal Society, when he and his postdoc Rick Wood made the first breakthrough towards identifying the UV-repair enzymes defective in xeroderma pigmentosum. Tomas promoted Rick into an independent position, and in 1995, Rick's lab purified and reconstituted in vitro the human enzymes responsible for nucleotide excision repair. This tremendous feat gave Clare Hall the double in the excision repair biochemistry championships, because in 1994 and 1996, Tomas's lab completed the equally laborious task of reconstituting base excision repair in vitro, using, respectively, the *E. coli* and human enzymes, some of which were remarkably conserved across the yawning evolutionary gulf.

In 2010, in recognition of his seminal contributions to the understanding of the biochemistry of DNA repair, Tomas was awarded the Royal Society's Copley Medal, the oldest scientific prize in the world, whose other recipients include Michael Faraday, Charles Darwin, and, latterly, Stephen Hawking. However, by this time, the other inhabitants of Clare Hall were rapidly catching him up in the scientific eminence stakes. Two in particular, Steve West and John Diffley, have used the enlightened environment of Clare Hall to undertake ambitious, long-term projects, the anathema of the normal three-to-five-year grant cycle. Their stories, which complete this chapter, illustrate everything good about Tomas and the lab he built with such success.

Steve West and Genetic Recombination

The Clare Hall Laboratories of the Imperial Cancer Research Fund opened at the end of 1985, with a small but star-studded cast of five lab heads. Tomas was joined by Jeff Williams, the only one of the Mill Hill developmental biologists who had not wanted to move to Oxford, and new recruits David and Birgit Lane, who had been working at Imperial College and Lincoln's Inn Fields, respectively. Jeff, David, and Birgit were well connected and greatly respected in their fields, and combined with Tomas's reputation, their presence contributed greatly to the sense that Clare Hall was going to be something good. Furthermore, even though they were not really within Tomas's original remit of DNA metabolism, they brought in some molecular biology skills that they were more than willing to teach to the biochemists.

Steve West, the last group leader to make up the new Clare Hall intake, had not intended to like Clare Hall at all, having only come for an interview from his then workplace, Yale, because he fancied a free trip to visit his mum in Yorkshire. However, almost three decades on, he is still an inmate, one of the big beasts of the DNA repair world, and probably Tomas's most inspired hiring. As well as all the normal attributes that a great scientist requires, Steve's reputation rests on his eye for major problems, his extraordinary skill as a biochemist, his phenomenal workrate, and his tenacity. All of these are exemplified in a knotty problem that Steve brought back with him from the United States, and that in the end took 25 years to solve: how does the cell cut through the tangled products of genetic recombination to regenerate two functional double helices?

Until the 1960s, genetic recombination was viewed by most biologists as an esoteric problem worked on by a small number of fungal geneticists. The exchange of information between two strands of DNA is essential for evolution and happens every time two people get together to reproduce, but pre–Watson and Crick, its study was caught in an intimidating briar patch of statistical analysis and complicated hypotheses. However, like everything else in biology, the DNA revolution meant that the geneticists' ideas could be reformulated in terms of the interaction of DNA molecules, opening the way for biochemists and molecular biologists to start thinking about exactly how recombination worked.

Recombination involves the breaking of a DNA double helix, the invasion of one of the strands of the broken helix into a homologous (identical) sequence, generally on the other partner chromosome, a period in which both chromosomes are joined together in a complicated four-way knot called a Holliday junction, and the cutting of the knot to resolve the tangle into two discrete double helices. Before resolution, the Holliday junction can move along the DNA strands, creating a situation in which, as if in an elaborate barn dance, strands can switch partners, ending up, when the molecular music stops, on the other chromosome from which they started. This process of branch migration, during which genes or parts of genes cross over from one parent's chromosomes into the other's, creates the concoction of mixed-up inheritance known on a small and delightful scale as a new baby, and on a larger scale as the basis of genetic diversity.

Early on in the history of recombination research, it became evident, from work performed by Dick Boyce and Paul Howard-Flanders, the codiscoverers of excision repair, that homologous recombination was also very important for repairing DNA damage. The most dangerous hit a DNA molecule can take is one that breaks both its strands simultaneously, because there is then no template from which to repair the lesion. Like a snapped rope, the flailing strands can be spliced to anything else that happens to be around, or simply left dangling, and the consequences to the cell can be disastrous: Large- and small-scale mutation and problems with replication lead inexorably to the catastrophes of cell death or (in multicellular organisms) cancer. Homologous recombination is one of two mechanisms for fixing double-strand breaks and averting these unpleasant fates.

In 1965, Alvin Clark and Ann Dee Margulies, working in Berkeley, published a set of *E. coli* mutants defective for both recombination and repair of UV damage, one of which they named *RecA*. By the mid-1970s, it had become clear that whatever the RecA protein did, it was central to both recombination and repair. A great deal of effort was expended to discover its identity, culminating in 1977 with the publication of three papers detailing the purification of RecA protein. One of this trio was an early appearance in print of Steve West, then a young lad from Hessle, near Hull, doing a PhD whilst supplementing his income DJ-ing round the student haunts of Newcastle.

Steve West's dad was a fish buyer in Hull, and despite parental expectations, Steve knew very early on that he did not wish to follow his father into a job at the fish docks: "He had a big red face, typical of someone who works outside in the cold, and his hands were … well, you can imagine! I can remember when he came home he stank of fish, and had fish scales all over him, and I thought, 'Nooo, that's not for me.' " Instead, Steve became the first in his family to go on to higher education, spending three years at Newcastle University to earn an undistinguished degree in Biochemistry: "I had too many other interests, and I could never be bothered to learn all the stuff, so I was never very good at exams. I was absolutely crap at maths too." Despite his unpromising record on paper, Steve got a PhD place at Newcastle with Peter Emmerson, an ex-postdoc of Paul Howard-Flanders, and realised that finally, he had found something that fired his enthusiasm: "I got in the lab and life was easy! You don't have to learn anything, and I always knew what the next thing to do was. I was very good at the bench and I was willing to put myself out to do experiments that would take a long time. I was obsessive about it, discovered what I liked doing."

Steve's project was to work on a mystery protein, unoriginally called Protein X, which appeared in vast quantities following irradiation of *E. coli*. It was not a subject of great interest to his supervisor: "Peter was famous for saying: 'Protein X is for the birds—why are you bothering with this stupid thing?' " but Steve hung on, reckoning that if a cell made something in such large amounts, it had to be important. He was more than vindicated when he discovered that Protein X was in fact RecA, and the 1977 paper describing this finding set Steve well on his way in the world of DNA metabolism. He moved to Yale to the lab of Paul Howard-Flanders, who turned out to be an understated Englishman with an almost pathological disinclination to dictate his new postdoc's research direction. Under the circumstances, the easiest path for Steve was to simply continue with his RecA work, which in any case was still very interesting, so this he duly did.

Life at Yale was very happy for a single-minded biochemist: "I stayed all night, worked fifteen hours a day, seven days a week. I used to take my girlfriend to a Greek restaurant near the lab, and the nice thing was, I could nip out between courses and change dialysis buffers, and she could stay there and go for a dance—it was one of those little restaurants where they do Greek dancing—until I came back from changing the buffer. She thought the mad scientist thing was cute." Much to Steve's good fortune, this amazingly tolerant person eventually agreed to marry him.

Steve published 20 papers during his seven years at Yale, in the process converting his boss, who had started life as a radiobiologist, into a confirmed believer in biochemistry. The RecA field was incredibly competitive, with several labs, including another one down the corridor at Yale, trying to figure out what RecA did during recombination, but at the tail end of 1980, Steve hit the jackpot. He was looking at what happened when purified RecA was mixed with two separate DNA molecules, which he was able to distinguish because he had tagged one with radioactive tritium and the other with ^{32}P. Instead of seeing changes in the

Steve West, 1978.
(Photograph courtesy of Steve West.)

sizes of the DNA bands, or perhaps the appearance of a new one, Steve got a most peculiar result; the tritium and ^{32}P labels seemed to have got mixed up between the two DNAs, so that both fragments contained both labels. After a couple of repeat attempts, during which the mixed up labeling stubbornly persisted, Steve set the whole experiment up again on Christmas Eve. He relates what happened next:

> I came in on Christmas morning, and half the tritium was still in the ^{32}P, and half the ^{32}P was still in the tritium. So I called Paul at home at lunchtime on Christmas Day, and I said: "I've got this really weird result and I just don't understand it." ... And there was this absolute silence on the phone, like a minute. Paul was actually a little strange in that he would have these speech gaps, he would stop talking for some time, but this was abnormally long. So I'm sitting there, and a minute goes by, and he said: "Well Steve, if it's a recombination protein, maybe that's what it's supposed to do." And at that point I realised it had done strand exchange! And then we were both kind of silent to each other on the phone, because it had done strand exchange—it had done something which was miraculous. And that was cool—a cool moment.

RecA was, indeed, the catalyst for strand exchange, coating the loose end of a DNA molecule to create a nucleoprotein filament, searching around for a homologous partner, pushing its double helix apart, and feeding the broken DNA in to start the process of recombination. For the next three years, Steve pretty much published whatever experiment he did, as the field leapt forward, armed with the knowledge of RecA's function. By 1983, Steve had worked out conditions in which he could take his purified RecA, add it to two different plasmids in vitro, and initiate strand exchange. Then, by adding some DNA polymerase and DNA ligase (the enzyme that glues loose ends of DNA together), he could magically transform the two circular plasmids into a single, figure-of-eight molecule joined at its waist by a Holliday junction. If put back into an *E. coli* strain that lacked the *RecA* gene, such figure-of-eights were resolved into a mixture of products that could only have arisen by homologous recombination. This elegant experiment, published in *Cell*, unequivocally showed that Holliday junctions were really the recombination intermediates, which had been suspected but not definitively proved, and also showed that RecA was not required for the final junction resolution step in the recombination reaction.

At this point, having been at Yale for five years, Steve knew it was time to go on the job market, for which he needed an interesting project that was entirely his own. The mechanism of Holliday junction resolution was a complete black box, so, being nothing if not ambitious, Steve decided that he would go hunting for the enzymes involved, with the goal of getting the whole of homologous recombination to work in vitro. In his last two years in Paul's lab, he set up the basics for the resolvase project, doing the lab three-week preps to make figure-of-eights, cracking open bacteria to make protein extracts, and seeing if any of the extracts could resolve the figure-of-eights in vitro. Much to his credit, Paul funded all of the work, but wanted none of the glory, a practice that Steve maintains: "He was fabulously generous and I try to do the same with people in the last year they're with me—do what you want, I don't care, and I don't want to be on the paper. And it's good, it allows them to have their own independence."

Despite having some good job offers from places in the States, Steve was seduced into coming to Clare Hall after meeting Tomas at an end-of-conference banquet in Cambridge:

> It was a nice evening and Tomas likes his wine, so I think I drunk far, far too much! A couple of weeks after I got back he sent me a letter which said: "We're building this place at Clare Hall, and would you like to come and

> have a look at it, and see what we have to offer?" I actually had no interest at all, because I'd been in the States for seven years, my girlfriend, who became my wife, was American, and we had no intention of coming back here. But my mother still lived here, so I thought, free trip back, go and see Mum. So I visited my Mum, and then came to Clare Hall, which was still a hard hat site. Tomas was pretty astonishing. He said: "If you come here we'll give you the resources—a technician, a graduate student, a couple of postdocs after a year or so, you don't have to write grants, and you can decide where you want the benches to go, where you want the sinks to be." He actually said those golden words: "If you can't succeed here, you probably won't be able to succeed anywhere," which was very clever actually, because when you're starting a lab you're scared to death you'll fail.

Steve duly arrived in his empty lab space in 1985 and got on with his resolvase-hunting experiments. Things had become a little easier since he'd begun them, because the three-week figure-of-eight preps had vanished into the past, in favour of sticking together four radioactively labeled artificial oligonucleotides (short lengths of DNA, in this case, 60 nucleotides long) to create a cruciform structure that was a perfect in vitro mimic of a Holliday junction. With this innovation, introduced by Nev Kallenbach at the University of Pennsylvania in Philadelphia, experiment times were reduced to afternoons, rather than weeks, and you could tell whether the junction had been cut simply by running the reaction products on a gel and seeing whether any of the original oligonucleotides had changed size from 60 bases long to 30. By 1989, Steve and his first postdoc, Bernadette Connolly, could definitely get resolution of the cruciforms by adding *E. coli* extracts to them, but try as they might, they simply could not purify the responsible enzyme from the bacterial protein soup. At this point, having exhausted the heavy guns of biochemistry, Steve decided to deploy the more subtle weapon of genetics, calling up his mate Bob Lloyd, who was one of the best geneticists in the field. Steve and Bernadette had already dabbled with making protein from known bacterial recombination mutants, to see if any of them were defective in resolvase activity, but all had been able to resolve cruciforms with no problem, showing that they must still contain functional resolvase. However, Bob Lloyd had four new mutant *E. coli* strains—*ruvA*, *ruvB*, *ruvC*, and *orf26*—which he was happy to send down from his lab in Nottingham to Clare Hall.

In the last experiment he did himself before retiring from the bench, Steve made cruciforms, added extracts from each of Bob's mutant strains, and ran the diagnostic gels. Using normal, *ruvA*, *ruvB*, and *orf26* extracts, he got the expected 30-nucleotide-long resolution product, but with the *ruvC* extract, there was a complete blank at the expected place on the gel. It looked as though *ruvC* might be the elusive resolvase, and this was rapidly confirmed by getting a *ruvC*-overexpressing strain from Bob Lloyd and showing that it was superefficient at cruciform resolution. In 1991, the West and Lloyd labs published a *Nature* paper describing the purification of the RuvC resolvase, which they showed was a nuclease able to sit on Holliday junctions and cut two opposing DNA strands symmetrically.

Steve's lab was now on a roll and showed very quickly that the RuvA and RuvB proteins were responsible for branch migration, acting together as a molecular motor that could move Holliday junctions along DNA until the RuvC protein, squatting on top of them, found a sequence that it liked, and cut the DNA to resolve the junction. Thanks to his work, and that of a few other labs, notably that of Hideo Shinagawa in Osaka, the mechanics of homologous recombination in bacteria were pretty much sorted out, and not a moment too soon; research into DNA repair and homologous recombination was about to become extremely trendy.

The sudden upturn in interest came about after two key findings. Firstly, the machinery of DNA repair and recombination was shown to be evolutionarily conserved to a remarkable degree, such that close relatives of many of the bacterial enzymes were found in mammals, and, secondly, mutations in the repair machinery started to crop up in cancers, beginning with the discovery in 1993 that a form of hereditary colon cancer was caused by defects in the repair of mismatched nucleotides. The tentative connection between DNA repair and human health, first made by Jim Cleaver in his xeroderma work, had finally clicked firmly and irrevocably into place.

Steve, somewhat unwillingly, was dragged out of bacteria into mammalian cells, although the Clare Hall facilities made the transition less alarming:

> It was postdoc pressure! Postdocs would join the lab, and I don't think they saw their future in bacterial proteins—they wanted to go into mammalian systems and find the same things. So the lab evolved from bacteria into human cells, probably over two years. The beauty was having Cell Services here—rather than growing *E. coli*, they would grow us mammalian cells, and we were just making extracts, so what's

> the difference? Extracts are extracts and you do the same experiments. So the transition was invisible—there was no real decision to be made, and the lab evolved from working on bacteria to human systems. Mind you, it would never have happened without Cell Services, because I've never grown a cell, and if I had a lab in a university, and had to grow cells in my own lab, I might have never made that transition.

As well as making a start on the search for the mammalian version of the RuvC resolvase, Steve had not forgotten his first love, RecA. In 1994, his lab was the first to purify its human homologue, RAD51. RAD51 coats single-stranded DNA to create a nucleoprotein filament that invades a homologous double helix, starting the process of recombination in a very similar way to RecA. The subsequent discovery that RAD51 was both positively and negatively regulated by the BRCA2 protein in response to DNA damage brought this work to centre stage in cancer biology. BRCA2, like p53, is a tumour suppressor, essential for guarding the cell from possibly oncogenic mutations. Without it, homologous recombination is defective, and double-strand breaks can only be repaired by a more error-prone mechanism called nonhomologous end-joining, leaving the cell extremely vulnerable; that mutation of *BRCA2* is found in some 10% of inherited breast cancers is therefore no surprise.

As the West lab grew and diversified, they continued to take on and meet multiple biochemical challenges in mammalian cells, all relating to defects in DNA repair and their effects on disease. They worked out that the huge BRCA2 protein, a nightmare to handle because it is so big and likes to fall apart, controls the ability of RAD51 to bind DNA, and they sit at the forefront of research into repair-linked diseases such as Bloom's syndrome, Fanconi Anaemia, and the unpleasant progressive neurological condition Ataxia with Oculomotor Apraxia 1. For all these achievements, Steve has won many awards and has been a fellow of the Royal Society since 1995. However, through all the years of success, Steve carried another project close to his heart, a failure-ridden enterprise that, despite its promising start, was foundering in a sea of dashed hopes and postdoctoral frustration: the quest for the mammalian resolvase.

To begin with, things had gone very well. As early as 1990, Steve's first graduate student, Kieran Elborough, was able to detect resolvase activity in extracts from mammalian cells, but as the decade wore on and the *Cell*, *Nature*, and *Science* papers rolled in for everything else, the resolvase project stuck fast. The problem was in the purification: "If you purified it, you'd lose activity, and then the strangest thing in the world would happen—the activity would come back. And we know now it's all to do with post-translational modification—it's inactive when it's phosphorylated; it's activated by dephosphorylation. So it would be coming off a column with other things that would phosphorylate and deactivate it. You'd take that, fractionate it again and lose the phosphates, and the activity would come back again. It was so frustrating. SO frustrating!"

The century turned over, and Steve was elected to the Academy of Medical Sciences and won the Leeuwenhoek Prize of the Royal Society, capping that in 2008 with Europe's most prestigious scientific award, the Louis-Jeantet Prize for Medicine. The resolvase project still languished in the doldrums and had even lost its sad little entry in the London Research Institute Annual Report. The project was so unsuccessful and such hard work for people that it gained an unhealthy reputation as a postdoctoral graveyard. In the end, Steve had to let it go fallow for two or three years because he simply couldn't talk anybody into trying it.

By 2005, Steve was getting pretty desperate: "It was funny, people said to me: 'What if after all these years somebody else finds it?' and I said: 'Fantastic! I just want to know what it is now, I don't care!' " However, salvation was at hand, in the shape of Stephen Ip, a new postdoc from Hong Kong who was even fonder of hard work than Steve himself and was, in addition, willing to take on a risky project if his supervisor insisted. Together with a second postdoc, Uli Rass, Steve and Ip formulated a last-ditch, two-pronged assault on the resolvase problem. Uli, subsequently joined by Miguel Blanco, would perform a massive screen using a molecular library containing every gene from the yeast *Saccharomyces cerevisiae.* The genes all came attached to a TAP-tag, a clever modification by which the proteins encoded by the genes could be easily hooked out of cell extracts. The plan was to individually purify all of the yeast proteins with unknown functions and test them one by one for resolvase activity using cruciform substrates. There were about 2000 to work through. While this was going on, Ip would make extracts from 200 L of the human cancer cell line HeLa; fractionate the extracts by protein size, charge, and chemical behaviour; and test the individual fractions for resolvase activity in the cruciform assay. Once the fraction with the most activity had been identified, the proteins in the fraction would be further separated according to size by running on gels. The size-separated samples would be tested again for resolvase activity, and, finally, the proteins in the sample with maximum resolvase activity would be identified by their mass-to-charge signature on a mass spectrometer. The whole process was as expensive as it was laborious, and no sane grants committee would have sanctioned it, so it was just as well that the Clare Hall core funding from Cancer Research UK was picking up the tab.

Uli slowly worked his way through the TAP-tagged yeast library, whilst Ip got on with the brute-force purification. And, finally, after 1100 TAP-tagged protein assays and Ip spending so long in the cold room running samples that he could have walked to the Antarctic and back, there came an afternoon in 2008 when Steve looked up from his computer and saw his heroic resolvase postdocs framed in the doorway of his office: "Uli, Ip and Miguel stood at my door and there were tears running down Ip's cheeks, because they knew they had it!" Almost simultaneously, Uli had isolated the yeast resolvase, Yen1, and Ip had found the human one, GEN1. Amazingly, they were homologous to each other but were entirely unrelated to RuvC, their bacterial counterpart. The 20-year journey from *E. coli* to humans had ended at last.

There would be many more twists and turns to come in the resolvase story, but for Steve, finding Yen1 and GEN1 was one of the biggest thrills of his career, easily up there with his Christmas morning realisation that RecA could do strand exchange: "It was a fantastic little story. It's fantastic just in terms of the ICRF and Cancer Research UK, because you could not have done this anywhere else in the world. You could not find any other funders who would fund the same project for twenty years. You had to be able to do this sort of thing without justifying it—it was at the point where it was becoming unjustifiable just to keep doing it."

The three resolvase postdocs. *(Photograph courtesy of Steve West.)*

And the future? "Now we know there are three to four pathways of resolution inside eukaryotic cells, of which GEN1 is one. It's fascinating because it turns out that the resolvases are controlled by post-translational modifications. So each is active at a different time in the cell cycle. Understanding what these proteins do, and how they are regulated and interact, should keep me going till I pack it in. I hope so, anyway!"

John Diffley and DNA Replication

In 1990, David and Birgit Lane were lured away from Clare Hall by the University of Dundee, giving Tomas an opportunity to hire new faculty, but also a problem. The loss of the Lanes triggered an outbreak of mingled pessimism and *Schadenfreude* in some parts of the scientific community; how would such a small, quiet place recover from the departure of two of its biggest guns? The notion that Clare Hall was doomed was soon scotched, however, by the surprise announcement that one of biochemistry's most visible and popular figures was moving there. Tim Hunt, who, most people thought, would only leave his beloved Cambridge when carried out feet first in a wooden box, had decided to abandon the creaky environs of the university Biochemistry Department for somewhere more congenial and better equipped, where he could concentrate on research rather than teaching. Tomas was delighted with his coup: "I was obviously very happy to take on somebody with Tim's high reputation. It gave me some personal satisfaction when people said to me, 'Oh dear, who are you going to find to replace David Lane?' and I was able to say 'Tim Hunt!' That really shut them up—none of them expected that answer!" Tim was equally pleased with his decision: "It was just wonderful suddenly to be surrounded by all these biochemists. Steve West was a total revelation to me when I went there. I was amazed—here was this guy putting recombination on the biochemical map. Sydney [Brenner] was always interested in recombination, but it was always phage genetics and that sort of thing—very very hard to understand."

After Tim's arrival, his contagious enthusiasm and boundless capacity for discussing even the most eclectic scientific subjects became one of the most energising aspects of Clare Hall life. When, much to his shock and amazement, he was awarded his 2001 Nobel Prize for the discovery of cyclins, the key regulators of the cell cycle, Tim's presence added a touch of star quality as well. His office became a high point of Frank Fitzjohns's Clare Hall supporters' tours, as Brenda Marriott recalls:

Tim Hunt and daughter, Nobel Prize ceremony, 2001, Stockholm. *(Photograph courtesy of Tim Hunt.)*

"Tim was wonderful with the appeals department. Frank would take the tours round that way and Tim would come to the door and stand there and smile, and after Frank told the tours who he was, he'd say very grandly, 'You can touch me if you wish!' And they would all start giggling!"

To complement Tim's interest in the cell cycle and to steer Clare Hall closer to his original ideal of a pride of DNA-focused lions, Tomas's other recruit, John Diffley, came from Bruce Stillman's lab at Cold Spring Harbor, which remains one of the major centres for research into the biochemistry of DNA replication. John is from Brooklyn, although he spent his school years in Suffern, New Jersey, where he attended a single-sex Catholic high school. At 16, his life changed radically when he got into New York University (NYU) as a premed. According to John, moving from New Jersey to university digs in Greenwich Village was "great, for all the wrong reasons and all the right reasons," but it didn't take long for him to realise that although New York was a great place to be, he had no desire to be a medic there or anywhere else: "During the first summer I worked in a hospital and realised I never ever wanted to do medicine—I hated it!" Instead, John enrolled at the age of 20 in the PhD programme at NYU to work with a junior group leader, Chris Brakel, who was interested in DNA replication. After a year, however, disaster struck: Brakel decided that she was going to leave science, and John was left high and dry, in possession of the last year of Brakel's grant money, but without a PhD supervisor. Undaunted, he wangled himself into another lab at NYU belonging to Mike Kambysellis, a Greek *Drosophila* geneticist, splurged the remnants of the Brakel grant on enough biochemistry supplies to keep himself afloat for the duration, and got on with his original project, purifying the main *Drosophila* replicative DNA polymerase. It was a crash course in independent research; nobody else was doing any biochemistry in the Kambysellis lab, and to cap it all, the lingua franca was Greek—"they'd have these massive Greek arguments—quite amusing!" So John was pretty isolated: "I never went to a meeting or anything. There was nobody to tell me to do this or that or to give me any guidance. I had to do it on my own, figure out what

I wanted to do. But I loved it; I completely loved it. I can't think of a better thing to do. I guess biochemistry is really puzzle solving—what most people love about science—figuring out how something works when you don't know."

Having had to figure out his next career move by himself as well, John got into Bruce Stillman's lab on the strength of the two extremely competent papers on polymerase purification that he published during his PhD. Any normal person, especially one coming out of John's situation, would probably have taken advantage of the new lab's expertise and done a solid biochemical project on DNA replication. However, John is nothing if not unusual: "I started doing yeast genetics, and again I taught myself. Bruce is a great biochemist but he doesn't do yeast, so I set up yeast in the lab. Everyone else was doing SV40 replication and I was doing yeast—I guess there's just a level of contrariness in my personality!"

Unsurprisingly, in the 30-odd years since Kornberg had found the first DNA polymerase, an awful lot had been discovered about the enzymes responsible for replicating DNA: Helicases unwound the DNA double-helix; primases started the new strands off; multiple types of polymerase, tailored to particular tasks, and each equipped with an editing function to cut out mismatches, got on with elongation; sliding clamps held the replicating strands of DNA steady; and ligases sewed the new pieces together to make the final product. The whole process resembled a busy factory production line, and although there was still considerable interest in the individual components of the process, attention was shifting to finding out how it was regulated. What was responsible for triggering replication, and how does the cell know to switch the replication machinery on once, and only once, during the cell cycle?

The Stillman lab was almost alone at the time in studying initiation, the first step in the replication process, in eukaryotic cells. As John Cairns had shown in the 1960s, bacteria, with their small genomes, start replicating from one particular point in their DNA, known as the origin of replication, and just continue until they have gone full circle. This is a relatively simple process. Eukaryotes, however, have a far larger complement of DNA, which is broken up into separate chromosomes, and therefore they need multiple origins of replication in order to perform the process on each chromosome, and within a reasonable timescale; in the case of mammalian genomes, origins of replication number in the tens of thousands. The problem that John wanted to tackle was how all

John Diffley, ca. 2010. *(Photograph courtesy of CRUK London Research Institute Archives.)*

these origins were able to fire just once in every cell cycle to orchestrate the complete replication of the genome, and, to begin with, he needed to define how the origins were recognised by the replicative machinery. During his time with Bruce, he purified two proteins that bound to yeast replication origins, and it was on the back of this, and other work in the lab's adenovirus model systems, that Tomas hired him.

Like Steve West before him, John applied to Clare Hall for less than scientific reasons; he fancied a trip to England, where he'd never been, and the ICRF job advert he'd seen in the back of *Nature* would pay his airfare. However, once he got here, the Clare Hall magic took over. Tomas remembers the stringent recruitment process very well: "I think I took him to the local pub and it was Sunday afternoon, and he was sitting looking out at the slopes of the ridge up there, a very nice British early summer day." John met the other group leaders, liked them, liked their work, liked the extremely generous starting offer, and decided to take the job: "The similarity to Cold Spring Harbor in terms of isolation is what appealed to me when I came here. I saw a lot of the same things in it and it either suits you or it doesn't. I did miss the people passing through Cold Spring Harbor at first—I started having to read the literature again! When you're at Cold Spring Harbor you don't have to read anything as you find out about it."

Following John's arrival, Tomas was slightly dismayed to find that his new recruit had, temporarily at least, abandoned biochemistry. Taking his example from the transcription field, where people were having a lot of success in reconstituting transcription in vitro by simply adding cell extracts to DNA templates, John had spent the end of his postdoc trying to do the same with replication but had completely failed. It was a perfectly sound strategy, but what he didn't know at the time was that replication is so tightly controlled that most of the proteins involved are either absent or held in an inactive state until immediately before they're required; adding random extracts was never going to work. In the face of this defeat, John decided to develop some approaches to look at what was sitting on the yeast origins of replication in vivo, in the cell itself, and for this, he had to use an extremely fiddly technique called genomic footprinting.

The principle of genomic footprinting is simple: if proteins are bound to DNA, the DNA is protected from attack by nucleases. Therefore, if you open cells up very gently, making sure not to disrupt the protein–DNA interactions, and then use nucleases to digest the DNA, the place where your protein of interest is bound shows up as a larger, undigested fragment of DNA. This can be visualised by the usual method of adding huge amounts of radioactive ^{32}P to label up the DNA, and running it on a gel afterwards. In practice, the experiment is extremely difficult to perform cleanly; it's really fiddly, with multiple steps, so there are many opportunities to lose material, which is a big problem because you are dealing with vanishingly small amounts of DNA in the first place. Furthermore, at the end of all your labours, you are often confronted with a complete mess, with so many background bands on the gel that the footprint is uninterpretable. It took John and his PhD student Julie Cocker almost three years to troubleshoot, during which time Tomas got progressively more worried: "It was clearly a very difficult project, and it was clear he wasn't getting anywhere with it. So I had a talk with him to see if he was sure it was what he should be doing. He felt the solution was just around the corner and he was right. But I did pay attention that way if somebody was trying something that was too ambitious."

The result of John's departure from biochemistry was a *Nature* paper in 1992, identifying a complex, called the origin recognition complex, or ORC, that was bound to yeast origins of replication in real time in the cell, as well as in a test tube. However, there was a lot more to be done. Although ORC was clearly sitting on origins of replication, it was unlikely to be what determined when those origins fired during the cell cycle; it was bound to DNA almost all of the time, and the protein that flicked the "on" switch was likely only to be there right before it was needed.

John's lab expanded their footprinting operations and found that one particular DNA sequence in yeast replication origins was only occupied by proteins during a phase of the cell cycle called G_1, which occurs immediately before S phase, the point where the genomic DNA is replicated. The new data appeared in a major paper in *Cell* that showed that this protein complex, dubbed the Pre-Replicative Complex (preRC), was very similar to *Xenopus* Licensing Factor, an activity in frog eggs that was required for initiation of replication. That similar complexes were found in two such dissimilar organisms suggested that the control of replication regulation might be broadly the same in all eukaryotes, and this has, indeed, turned out to be the case.

With the genomic footprinting working well, John deployed his other nonbiochemical skill, yeast genetics, to identify mutants unable to assemble the preRC properly. Using these mutants, he could bootstrap his way to the proteins they encoded and start purifying them, with the aim of reconstituting the preRC in vitro, finally returning to his roots as a biochemist. John had already started a very successful collaboration on the ORC with the yeast molecular biologist Kim Nasmyth, and in 1996, the two labs published a joint *Nature* paper, identifying the first preRC complex protein, Cdc6.

The preRC is the thread running through all of John's subsequent research, and 18 years on from the first in vivo footprint, John's lab finally achieved its in vitro reconstitution. This scientific and technological tour de force put a vital piece into the puzzle of how replication is limited to a one-time event during the cell cycle, because regulation of the preRC turns out to be the most important thing. The cell has evolved an elegant system for preventing origins from firing more than once: only allow the preRC to be assembled at a particular time in the cell cycle, and only allow its subsequent activation during S phase if the cell is confident that all systems are go.

The first step, preRC binding, now called licensing, is regulated so that it can only occur in the early stages of the cell cycle, from just after the cell has split in two (mitosis) until the end of G_1. The ORC, together with Cdc6 and another protein, Cdt1, uses the cellular energy source ATP to load a helicase motor, comprising another six proteins together known as Mcm2–7, onto origins. The preRC complexes sit on the origins and wait. Although they are ready to go, they need the biochemical equivalent of car keys to get their engines going, and these are only provided by the cell during S phase, in the shape of specific activating enzymes that are not made in any other stage of the cell cycle. The enzymes, which belong to a group called cyclin-dependent kinases (CDKs; see Chapter 6 for more about how they work), turn on the Mcm2–7 helicase, which starts to move along the DNA, unwinding it as it goes so that the rest of the replication machinery can load and begin synthesis. After the genome has been replicated, the cell divides, and the whole procedure starts again.

The processes of assembling the preRC on DNA and getting it activated are so crucial to correct replication that there is layer upon layer of regulation of both. Working out the intricacies of how the cell achieves its aim of perfect reproduction of the genome has taken John into the fields of transcription, whose enzymes have had to learn to share nicely because they use the same template as the replicative machinery, and, no doubt to Tomas's satisfaction, into DNA repair territory. To start DNA replication, the cell has to pass a checkpoint, where DNA damage is detected and reported to the replication machinery. A certain amount of careless, error-prone replication is allowed in the interests of keeping the S-phase show on the road, but if things look really bleak, the cell calls a halt to the whole process and initiates a frantic repair programme before allowing S phase to continue. Any origins that have not yet fired are inactivated by switching off the Mcm2–7 helicase, and the replication machinery sitting at stalled replication forks is stabilised, until it receives the signal to resume synthesis.

Such complexity of regulation might appear daunting, but to a biochemist of John's calibre and ambitions, the thought of purifying 60–70 proteins and then assembling them into somewhere around 25–30 complexes is his idea of the ultimate good time:

> This may be a bit of a boy thing, but ever since I was a little kid I took things apart. I broke every toy I ever got for Christmas within a week because I took it apart and couldn't put it back together. So now that I'm finally putting things back together, it's great! There are really interesting questions to ask about how these things work. What's particularly interesting is that there isn't a precedent in viruses or in bacteria—it really works completely differently from replication in any other realm of life as far as we know, so the things we're going to learn are really new. Replication is the centre of everything in the nucleus, because everything that happens is coupled somehow with replication, whether it's chromosome inheritance or sister chromatid cohesion, so I think having an in vitro replication system is really going to open up a lot of stuff for biochemical analysis.

In the short term, the loading of the Mcm2–7 helicase onto the origin is currently occupying John's thoughts because persuading a multisubunit helicase to wrap itself round DNA is an intriguing topological problem. In addition, his lab has just found that ORC, which together with its accessory protein Cdc6 is the ATP-burning enzyme that loads the helicase up, can tell whether the helicase is intact or not, and unload it if components are missing. The ORC ATPase, unsurprisingly, also has a further level of regulation, by CDKs.

In the long term, John's objective is simple: "We're close to having the whole replication process in vitro. We've got about a third of the proteins purified, but in yeast we know all the proteins involved and it really just

Nobel Prize celebrations, Clare Hall, 2001. (*Left* to *right*) Steve West, Frank Fitzjohn, Tim Hunt, Tomas Lindahl, Brenda Marriott. *(Photograph courtesy of Steve West.)*

means purifying all the proteins and reconstituting them. But my real goal is that I want to reconstitute a yeast mini-chromosome with replication and a gene, so we can look at interactions between replication and transcription. I think that's a feasible thing." His old boss, Bruce Stillman, agrees: "The extraordinary clarity of his observations, the crispness of his data, and the beautiful logic of his ideas and discoveries have been characteristics of John's research. He has always addressed important problems and has been at the forefront of solving them. I would venture to say that John is one of the best scientists today who focusses on how our genome is inherited from one cell generation to the next" (Stillman 2003).

Web Resources

www.nobelprize.org/nobel_prizes/medicine/laureates/2001/hunt-autobio.html#
Unfortunately for this author, Tim Hunt's Nobel Prize was won for work conducted in Cambridge and Wood's Hole and therefore falls outside the remit of this book. To read more about him and his work on cyclins, a good starting point is the Nobel website.

www.youtube.com/watch?v=lZxc3Pohf70 Good video about DNA structure.

www.youtube.com/watch?v=cvU4kEMvbm4 Fantastic video of Matt Meselson describing "the most beautiful experiment."

www.youtube.com/watch?v=SROcb5h1-w Useful animation of a Holliday junction.

Further Reading

Cairns J. 1997. *Matters of life and death.* Princeton University Press, Princeton, NJ.

An extremely readable account of the things John Cairns thinks are important about science, mingled with a bit of autobiography in passing.

Diffley JF. 2011. Quality control in the initiation of eukaryotic DNA replication. *Philos Trans R Soc Lond B Biol Sci* **366:** 3545–3553.

Friedberg E. 1997. *Correcting the Blueprint of Life: An historical account of the discovery of DNA repair mechanisms.* Cold Spring Harbor Laboratory Press, Cold Spring Harbor, NY.

I am indebted to this simultaneously clear, entertaining, and erudite account of the early days of DNA repair—well worth searching for.

Kornberg A. 2000. Ten commandments: Lessons from the enzymology of DNA replication. *J Bacteriol* **182:** 3613–3618.

Lindahl T. 1996. The Croonian Lecture 1996: Endogenous damage to DNA. *Philos Trans R Soc Lond B Biol Sci* **351:** 1529–1538.

West SC. 1996. The RuvABC proteins and Holliday junction processing in *Escherichia coli. J Bacteriol* **178:** 1237–1241.

Quotation Sources

Bruce Stillman remarks on John Diffley for Paul Marks Prize for Cancer Research 2003 Prize Winners, Memorial Sloan-Kettering Cancer Research. www.mskcc.org/research/paul-marks-prize-research/2003-prize-winners.

Cairns J. 1963. The bacterial chromosome and its manner of replication as seen by autoradiography. *J Mol Biol* **6:** 208–213.

Cairns J, Stent GS, Watson JD. 2007. *Phage and the origins of molecular biology: The centennial edition.* Cold Spring Harbor Laboratory Press, Cold Spring Harbor, NY.

Friedberg EC. 1997. *Correcting the blueprint of life: An historical account of the discovery of DNA repair mechanisms.* Cold Spring Harbor Laboratory Press, Cold Spring Harbor, NY.

Hanawalt PC, Hayes RH. 1967. The repair of DNA. *Sci Am* **216:** 36–43.

John Cairns Biography. Oral History Collection, Cold Spring Harbor Laboratory Archives. http://library.cshl.edu/oralhistory/speaker/john-cairns/.

Kornberg A. 1960. Biologic synthesis of deoxyribonucleic acid. *Science* **131:** 1503–1508.

Matthew Meselson Biography. Oral History Collection, Cold Spring Harbor Laboratory Archives. http://library.cshl.edu/oralhistory/speaker/matthew-meselson/.

Susman M. 1995. The Cold Spring Harbor Phage Course (1945–1970): A 50th anniversary remembrance. *Genetics* **139:** 1101–1106.

CHAPTER 5

Brake Failure

In the summer of 1952, around the time that Renato Dulbecco was kick-starting the animal virus field in his Caltech basement lab, one of his close friends and compatriots was on her way towards a discovery that would be equally momentous. Rita Levi-Montalcini, like Dulbecco an ex-student of Giuseppe Levi, the extraordinarily charismatic Professor of Anatomy in Turin, was boarding a flight from Rome to Rio de Janeiro, where she was to spend a few months in the lab of another of Levi's protégés, Hertha Meyer. She was not alone. In her pocket, safely stashed in a small cardboard box with an apple as an in-flight snack, were two white mice, who had already traveled to Italy with Rita from her lab at Washington University in St. Louis, Missouri. Unfortunately for the mice, they were not travelling as pets; both carried tumours, vital components of the work Rita hoped to do in Brazil.

Unlike her friend Renato, who had moved to the United States at the same time to learn molecular biology with another Levi alumnus, Salvador Luria, Rita had continued to work in Giuseppe Levi's own field of experimental neuroembryology. Her move to St. Louis in 1946 had been at Levi's prompting, to the lab of Viktor Hamburger, who had developed a system to study how the growth and migration of nerve fibres was directed by the nonneuronal tissue of embryos. The experimental model used in the Hamburger lab was the embryonic chick limb, a firm favourite of embryologists for both prosaic and scientific reasons: eggs were cheap, it was easy to watch what was going on in them by cutting a window in the shell, and the chick embryos were sufficiently large to make surgery such as grafting possible. Work in the lab was founded on the phenomenon that if the chick limb bud, the centre from which the limb grew, was completely destroyed, the nerves that would have grown into the limb also died. Conversely, if an extra limb bud was grafted onto an embryo, new nerve fibres would grow towards and enter the transplanted limb.

Things did not go well to begin with, and Rita, a frequent visitor to Dulbecco and Luria in Bloomington, Indiana, was tormented by doubts about whether she was doing the right thing. In her autobiography, *In Praise of Imperfection*, she relates that she had

> ... an unconfessed envy for geneticists and microbiologists, whose work I believed was far more interesting and promising than that of neuroembryologists.... I did not hide from ... Luria my distressing doubts about the validity of our approach to the problem of the differentiation of the nervous system. I believe his doubts were much greater than mine, but he saw no way out of the difficulty. He excluded the possibility of my successfully taking up microbiology, unprepared as I was in the field, and encouraged me, without enthusiasm, to continue in my research (Levi-Montalcini 1988).

Rita's frustration with her work lay in the sheer intractability of studying neuronal development; compared with the simplicity and solid theoretical grounding of molecular biology and genetics, it seemed a completely closed system. She could look at the growing nerves, by fixing and staining sections cut from the embryos, but there seemed no way of working out what was going on. However, in 1947, she made a conceptual breakthrough that completely restored her faith. Whilst examining sections cut from embryos at slightly different stages, she realised that she could track the temporal development of the nerve cells. As she says, "the startling realization that nerve cell populations were subject to quotas and to the elimination of excessive numbers in their ranks, as well as to migrations that went hand in hand with functional differentiation, showed that there were ontogenetic processes at work in the nervous system which were not as inaccessible to investigation as I had previously imagined" (Levi-Montalcini 1988).

In 1950, Hamburger received a letter from an ex-student of his, Elmer Bueker, which further galvanized Rita. Bueker was comparing the abilities of different mammalian tumours to grow as grafts on chick embryos, and had noticed that one tumour, S180, a mouse sarcoma, exerted an extraordinary effect on the nervous system of the host embryo. Nerve fibres originating from nearby sensory ganglia (clumps of nerve cell bodies) had grown into the tumour mass, completely innervating it just as if it were a grafted limb bud. Intrigued, Rita ordered in some S180-carrying mice and repeated Bueker's experiment. What she saw was quite astonishing. "The mass of tumoral cells ... were from all sides penetrated by bundles of brown-blackish nerve fibers. These fiber bundles passed between the cells like rivulets of water flowing steadily over a bed of stones.... I had the feeling that I had come upon a phenomenon without precedent in the rich case history of experimental embryology" (Levi-Montalcini 1988).

Rita repeated the experiment with a different mouse sarcoma, S37, and got the same results. Crucially, she then showed that even when she physically separated the tumours from the growing chick by grafting them outside the membrane encapsulating the embryo, the tumours still had the same effect. Rita realised to her great excitement that the tumour cells had to be releasing a soluble factor that was neurotropic, able to lure nerve fibres towards it, and that could also direct the future development of immature nerve cells, instructing them to differentiate into specific neuronal cell types. This factor, christened Nerve Growth Factor, soon abbreviated to NGF, turned out to be the key to the puzzle of how non-neuronal tissue could dictate how the nervous system developed.

Grafting tumours onto chick embryos was a cumbersome and slow technique, leading to Rita's decision to go to Rio with her two mice carrying the S180 and S37 tumour cells. She had been hoping for an excuse to visit her friend Hertha Meyer in Brazil for some time, and this seemed perfect. Hertha had set up neuronal tissue culture in her lab, and growing the tumours with neurons in vitro was a far cleaner way to study the mystery factor than continuing with chicken eggs. There were some initial setbacks, but by the time of her return to St. Louis in January 1953, Rita had established tissue culture conditions under which nerve cells would put out dense, halo-shaped outgrowths of fibres in the direction of tumour cells. In a salutary lesson to today's work-obsessed scientists, she also found the time to take a month's holiday, starting first with Carnival in Rio and then going on to Peru and Ecuador.

Although Rita had set up a good assay for NGF in Rio, the obvious next step, attempting to purify the factor, was quite daunting, requiring a set of biochemical skills that neither she nor Viktor Hamburger possessed. However, in Rita's absence, Viktor had made an inspired hiring decision—a young New Yorker called Stanley Cohen.

Stan Cohen had started his biochemical training on a PhD project that had probably not attracted a lot of applicants; in an attempt to study how the end product of nitrogen metabolism can be switched between urea and ammonia, he had had to spend many nights crawling around the lawns of Michigan for his experimental subject, the earthworm. Thousands of worms later, he graduated, armed with the unusual skill of being able to insert stomach tubes into tiny wriggly things, and had been hired by a lab in Colorado working with premature babies. In 1952, he arrived at Washington University as a postdoc in the Department of Radiology, but soon moved into Hamburger's lab, intrigued by the twin opportunities of isolating a novel protein and learning some experimental embryology along the way. He was exactly what Rita needed, and she was delighted with him:

> … immediately struck by Stan's absorbed expression, total disregard for appearances—as evinced by his motley attire—and modesty. He never mentioned his competence and extraordinary intuition which always guided him with infallible precision in the right direction.… He would arrive in the morning with a pipe in his mouth, limping slightly because he had had polio as a child.… He was followed by Smog, the sweetest and most mongrel dog I ever saw. Smog used to lie down at Stan's feet when he sat at his desk, and keep a loving eye on him, or slept when he fidgeted with test tubes or relaxed playing the flute.… I have often asked myself what lucky star caused our paths to cross. He, too, profitted from our collaboration. If I, in fact, knew nothing of biochemistry, Stan, when he joined us, had but vague notions about the nervous system. The complementarity of our competences gave us a good reason to rejoice instead of causing us inferiority complexes (Levi-Montalcini 1988).

Stan started off by repeating something Rita had already done in Brazil—grinding up tumours, making a crude extract from the cells, and adding this to neurons in tissue culture. The extracts worked very well, eliciting nerve fibres from sensory ganglia, but the source material was problematic. To work, the tumours had first to be implanted into chick embryos, where they grew as little nodules. To generate enough material for Stan to use, Rita was having to set up dozens of embryos for every experiment. After a year of hard slog, the two had shown that NGF was a protein complexed with some kind of nucleic acid, but getting further information about it, let alone purifying it, seemed a daunting task; a veritable mountain of chick embryos would have to be sacrificed to the cause.

At this point, Stan and Rita had an enormous stroke of luck. So as not to go completely native in the neuroembryological surroundings of the Hamburger lab, Stan was attending regular group meetings with some of the other biochemists at Washington University, who included that dragon of DNA replication, Arthur Kornberg. Kornberg suspected that the nucleic acid component of the NGF prep was, in fact, just a contaminant, and to get rid of it, asked one of his postdocs to give Stan some partially purified snake venom, the source of a nucleic acid–digesting enzyme called a phosphodiesterase. Stan set up an experiment in which he cultured neurons with partially purified NGF with or without snake venom. What happened next is related by Rita: "Among the fractions that I assayed in vitro the following day, there was one containing snake venom. Having not been told which of the fractions had been specially treated, I was completely stunned by the stupendous halo radiating from the ganglia. I called Stan in without telling him what I had seen. He looked through the microscope's eyepieces, lifted his head, cleaned his glasses which had fogged up, and looked again. 'Rita,' he said quietly, 'I'm afraid we've just used up all the good luck we're entitled to'" (Levi-Montalcini 1988).

The "stupendous halo" had been induced by the NGF prep containing snake venom, which, somewhat amazingly, contained NGF in a form that was far more potent than that found in tumours. From their new NGF source, Stan purified a fraction that was 3000 times richer in NGF, but there was a snag: snake venom was very expensive. Rather than bankrupting the Hamburger lab, Stan and Rita looked around for another NGF source. In an inspired bit of lateral thinking, Stan decided to try to make extracts from the closest object in the lab animal kingdom to snake venom glands, the salivary glands of male mice.

Energetic brawling is a way of life to male mice, and not surprisingly, post-fight recuperation involves a lot of wound licking. Back in the 1940s, a Frenchman, Antoine Lacassagne, had observed that the salivary glands of male mice were much bigger than those of the female, and Stan wondered whether the reason might be therapeutic; perhaps the saliva was enhanced with growth factors to aid post-fight wound healing. Very gratifyingly, he was right, because the glands turned out to contain a lot of an NGF that was very similar in structure and size to the snake version. The new NGF source meant that experiments could be scaled up dramatically, and in the winter of 1958, Rita started to look at what happened when she injected salivary NGF into newborn rodents. The effects were dramatic; the size of sympathetic ganglia increased 10-fold after only a few days' treatment, and there was also a big increase in innervation of internal organs and skin. Treating rodents with an antiserum able to block the activity of NGF had the opposite effect, selectively reducing sympathetic ganglia to near extinction, with little or no effect on any other part of the nervous system, or on normal growth.

At the end of 1958, Stan left the Hamburger lab for Vanderbilt, where he'd been offered the faculty position that budgetary constraints at Wash

U had denied him. Both he and Rita carried on with their work separately but never collaborated again. Rita mourned his departure: "I saw coming to an end the most productive period of my life and also the most picturesque in the saga of NGF, when Stan with his magical intuition and flute played the part of the wizard, charming snakes at will and getting the miraculous fluid to flow forth from the minuscule mouths of mice" (Levi-Montalcini 1988).

The work Stan and Rita had done, showing that a soluble factor was able to direct the development of the sympathetic nervous system of a growing organism, permanently changed the way scientists thought about how cells grow and differentiate. If it was true of one system, then it was likely that the multiplicity of cell types that together comprise a fully functional living creature were all being directed down the correct lineages by the actions of other, as yet unidentified, growth factors. Although such a model was inevitably too simplistic, growth factors did turn out to be some of the most important of a wider cast of players, and the NGF work opened up a whole field of research into how growth factors were made and how they signaled into their target cells. Not surprisingly, Stan Cohen and Rita Levi-Montalcini received the 1986 Nobel Prize in Physiology or Medicine for their discoveries. Sadly, Giuseppe Levi, in whose laboratory some of the biggest stories of biology in the 20th century originated, did not live to see any of his three Nobel protégés—Luria, Dulbecco, and Levi-Montalcini—get their awards. He died in 1965, although admittedly, he was 92 at the time so hadn't done badly. (Parenthetically, growth factor research seems to induce longevity in its practitioners. Stan Cohen is 90, Viktor Hamburger lived to the age of 100, and Rita Levi-Montalcini died in late 2012 at a magnificent 103.)

Despite considerable interest in the papers that Rita and Stan published in 1960 describing the new factor, there was little attempt to capitalise on their discovery for some years, mainly because of the technical issues involved in factor purification that Stan and Rita had already encountered. The twin challenges of finding an assay for an unknown growth factor and then identifying good sources of material from which to purify it were not trivial.

By the end of the 1970s, the biochemical approach to growth factor research was bearing fruit, but progress had been terribly slow. Stan and his lab at Vanderbilt had purified mouse salivary gland NGF in 1960, using around 75 mice for each 3 mg of protein, but it took another 11 years, until 1971, for the amino acid sequence of NGF to be determined by Ruth Hogue Angeletti and Ralph Bradshaw in St. Louis. Because protein sequencing, identifying the linear string of amino acids that makes up a protein, required very large quantities of pure starting material, Angeletti and Bradshaw had to use ~250 mg of NGF, still purified from male mouse salivary glands. Although their paper does not go into detail, they presumably had to use thousands of mice. A second factor, Epidermal Growth Factor, or EGF, was discovered by Stan Cohen as a contaminant of salivary gland NGF preps. Stan noticed that baby mice injected with partially pure NGF opened their eyes a week early because the skin of their eyelids was maturing unusually fast, and because it seemed unlikely that NGF would cause such an effect, he went hunting for the responsible protein. His laboratory purified and sequenced EGF in 1972 from the usual vast number of male mouse salivary glands.

In addition to EGF and NGF, several other labs managed to wholly or partially purify new activities, but these were all stuck in limbo as mystery proteins because none of them could be made in large enough amounts to satisfy the voracious sequencing machines. Although one could characterise the purified factors in terms of gross properties such as size and see what happened when they were added to cells or injected into animals, knowing their sequence would be the key to deciphering their real identity and functions. Protein sequencing was clearly the rate-limiting step in the whole process, and it would need a big improvement in the technology behind it to move the growth factor field forwards.

Protein sequencing had been around for a while. Fred Sanger of the Laboratory of Molecular Biology in Cambridge published the first complete protein sequence, that of insulin, in 1951, but long before he won his first Chemistry Nobel Prize in 1958 for this feat, his technique was superseded by the Swede Pehr Edman's eponymous Edman degradation method. Edman worked out conditions under which he could chop off and simultaneously label the first amino acid in a polypeptide chain whilst keeping the remaining chunk of protein intact; the products of the reaction were therefore a single amino acid, whose identity could be determined by various analytical methods, and the remaining peptide chain, which could be recovered and used for another reaction. Like cutting sausages from a string one by one, the Edman degradation could be repeated multiple times and was therefore much less wasteful than the Sanger method, which trashed the whole protein sample each time an amino acid was cleaved off and identified.

Edman and his assistant Geoff Begg spent the early 1960s turning the laborious manual method into something that could be automated. In 1967, Edman and Begg published their paper "A Protein Sequenator" in the *European Journal of Biochemistry*, together with the results of a sequencing reaction performed using 5 mg of humpback whale myoglobin, where they had identified 60 amino acids in a single run. At Edman's insistence, his spinning cup sequencer, as it was known, was not patented, and numerous DIY versions based on Edman's design were assembled, together with several commercially marketed machines, probably the best of which was sold by Beckman Instruments, Inc. Sequencing was now more accessible, but the first Edman machines, taking days to produce a sequence, were slower than manual sequencing and still required large amounts of protein. There was obviously a lot of scope for refinement, and many laboratories around the world took up the challenge.

Edgar Haber's lab at MGH in Boston was amongst the early adopters of the spinning cup sequencer, building a version that was a bit on the Wallace and Gromit side, but nevertheless worked quite well. Haber's research interests were in cardiology and immunology, but he was also an expert protein chemist, having spent three years in Chris Anfinsen's lab at the National Institutes of Health (NIH). Anfinsen and his colleagues, including Haber, were the first to show that the three-dimensional (3D) structure of a protein was entirely dependent on its amino acid sequence, for which discovery Anfinsen won a share of the 1972 Nobel Prize for Chemistry. By the mid-1960s, Haber was working on determining the structure of immunoglobulins, antibody molecules, which were then something of a mystery. The prevailing dogma, that one gene made one protein, was very hard to reconcile with the fact that the body could manufacture millions of different antibodies, more than could be encoded by the genome. Working out the structure and amino acid sequence of immunoglobulin molecules was clearly important, and in 1967, a young Englishman arrived in Haber's lab for a postdoc in this hottest of topics.

Mike Waterfield was one of the first graduates of Brunel College of Advanced Technology, a post-war innovation in higher education, whose Bachelor of Technology courses were almost unique in offering students industrial placements as part of their degrees. It was an interesting choice of college for a product of the highly academic Portsmouth Grammar School, but the adolescent Mike, despite or perhaps because of having a headmaster for a father, never worked particularly hard, and despite being the school chemistry star, lacked confidence in his ability to make it at a big university.

As it turned out, Brunel was a good fit, and the placements that Mike went on were a stimulating introduction to industry and academia, especially the six months he spent learning synthetic chemistry with Tom Connors at the Chester Beatty Labs in South Kensington. Connors, then only three years out from his PhD, was a sort of chemical Tigger, bursting with activity, both scientific and otherwise. As Mike recalls: "He was a legend. His favourite thing in life was chemistry, and his second favourite thing was practical jokes. We spent our time dropping bricks off the top of doors onto people and going to the local pub."

Having miraculously avoided any charges of brick-related GBH, after graduation in 1964, Mike started a PhD at Kings College Hospital Medical School, working with Charles Gray on porphyria, the disease responsible for the madness of George III. Somehow, crystallizing bilirubin and biliverdin from faeces was not sufficient to put Mike off bench work forever, and he became pretty good at protein chemistry and enzymology (and, no doubt, equally adept at holding his nose whilst doing experiments). In 1967, he left Kings, bound for America and the Haber lab, inspired by the explosion of molecular knowledge that characterised biomedical research in the 1960s.

In Boston, Mike, in addition to going on anti–Vietnam War marches, discovered that protein sequencing, rather than just being the means to an end, was, in fact, rather interesting in itself. He spent a great deal of his time in Haber's lab improving the instrumentation involved in analysing the succession of amino acid derivatives that came off the spinning cup sequencer. He and his colleagues played with different ways of detecting them, at one point going over to MIT to use their superior mass spectrometers. Unfortunately, although their contact there was happy to let them use the machines for a while, he eventually abandoned them in favour of a far sexier project: analysing the rocks brought back from the first Moon landing in 1969.

The work that Mike had done on developing new sequencing methodology was innovative enough that when he met Bill Dreyer of Caltech,

Dreyer was keen to import him into his own institute. Dreyer, Professor of Biology at Caltech from 1963 until his death in 2004, was part of the utterly remarkable generation of Caltech scientists that included Max Delbrück, Renato Dulbecco, Richard Feynman, Linus Pauling, and many others. He had an exceptionally well-developed visual gift that allowed him to think in 3D, building and testing virtual machinery in his head that he would then translate into real-life prototypes. In addition to his talent for instrumentation, Dreyer was also the first person to provide the correct answer to the question of how immunoglobulin molecules were constructed. He and Claude Bennett proposed in 1965 that they were made from split genes that could be spliced together in an almost infinite variety of ways. The idea was so radical that it took more than a decade and an awful lot of scepticism from the rest of the field before Susumu Tonegawa proved that they were correct (and in the process lined himself up for the 1987 Nobel Prize in Physiology or Medicine).

Rather than taking Mike on himself, Dreyer fixed him up with a place with Leroy Hood, his long-time protégé. Originally a medic, Lee Hood had been Dreyer's PhD student, but as an MD had been drafted as part of the Vietnam War effort to work in the Public Health Service at the NIH. Upon his release from duty, he returned to Caltech in 1970 to start an independent lab in the Biology Department. Hood and Dreyer shared an interest in instrumentation and methodology that was shortly to make Caltech the unrivaled world leader in sequence automation, but Hood, in Dreyer's words, was "a different kind of person. He's superb as a scientist, and he's superb as an administrator and fund-raiser, but he wasn't a risk taker, or an imaginative innovator, or whatever" (Dryer interview 1999). Mike concurs that Lee was principally an enthusiastic facilitator of the work going on in his lab: "He relied on the lab to do the work and we got on with it. We were able to do anything we wanted—there were never any money issues." Mike continued with his work on analytical mass spectrometry, doing a deal with the Caltech Jet Propulsion Labs to use a gold-coated mass spectrometer that was originally designed to go to Mars. Because the Hood lab's major biological interest was the structure of immunoglobulins, Mike also amassed a respectable set of papers on the immunoglobulin light chain, in the process making some major improvements to protein sequencing methodology.

Life at Caltech was pretty exhilarating. In addition to the amazing facilities and unlimited funding, the Biology Department was an interdisciplinary brainfest, full of opportunities to hear about new and exciting science from major labs around the world. Bill Dreyer ran a supercharged journal club once a week, featuring wine and intimidating questions about the papers everyone was expected to have read. Mike remembers it as being "very alarming. I definitely had to get used to being somebody who had to match up to these people. I didn't feel that I was in the same league, but I think the work was so exciting and we were recognised as being front edge people in what we were doing, so I guess it gave me some confidence."

Outside the lab, things were equally alluring. The music and club scene was amazing, possibly enhanced by the availability of extremely cheap marijuana, and there were also the traditional Caltech camping trips, although Mike's expeditions came with a little extra twist: "We spent nights in the desert building rockets using chemicals pinched from the chemistry department which we fired up into the sky. It was so dangerous, so crazy; we walked to the top of mountains to do it." There were more intellectual pastimes too—weekend lecture marathons on, for example, the history of art, and an art class shared with Richard Feynman, who took his love of life drawing a little further than Mike: "He used to go to all the strip clubs in Pasadena and practise drawing all the ladies there."

In 1971, Mike received an offer that would take him away from California on a new trajectory, when a Caltech alumnus showed up in the Biology Department laboratories. Mike Fried was on the lookout for a good protein chemist to fulfill Michael Stoker's desire to get the ICRF into the Machine Age, and someone of Mike Waterfield's calibre, reputation, and pedigree seemed the perfect choice. Mike also had the added advantage of being a cheap date: "Because of my connection with Ed Haber, I applied for a five year fellowship from the American Heart Association, which I got. I was the first person ever to be allowed to take that abroad. I waltzed back to the UK with an American salary at the fantastic exchange rate of the time, and with unlimited funding. I lived on dollars for five years—I didn't pay any tax (they caught me two years later!)."

Aside from the financial benefits, Mike took full advantage of both the scientific and personal aspects of living in London:

> Remarkable opportunities come from having freedom, critical mass, and people to inspire you, and Mike Fried took me back to an environment where there were a whole bunch of great people.... To present in room 401 was just nerve racking. Everything was flooding in enabling Mike and Beverly to clone and sequence polyoma. There was an influx of

> enzymology and technology and just a burst of massive information where people were excited, happy and hardworking. And the social life was … hmmm … very interesting! I found a cottage in the middle of Hampstead Heath—an old garage with rooms above it in a lane behind the Old Bull and Bush. There were lots of memorable events there—we held institute parties which were legendary. I used to bring the leaves in from the trees and have the floor a foot deep in leaves to dance in. I met Sally [Mike's wife, Sally James] there—I invited her to a party there and our relationship started dancing in the leaves to The Wrights, David Bowie and a few others from that era. That was the most important event in my life.

Before Mike's arrival in 1972, the most complicated machine at the ICRF was an ultracentrifuge, and not everybody was brave enough to use it. The presence of a protein sequencing expert, especially one as keen as Mike to collaborate with his colleagues, started the sea change in attitudes to new technology that underpinned much of the ICRF's subsequent upward trajectory. Because Bill House was running the funding with a fairly liberal hand, Mike was able to commission the building of a state-of-the-art protein sequencer at Caltech, which he further refined when it arrived in London, helped by Geoff Glayzer in the ICRF workshop. The close relationship with Caltech was to persist for the next decade, cemented by the emigration of Mike's technician and first PhD student, Rod Hewick, into Bill Dreyer's lab in 1978. (Hewick subsequently went on to sequence erythropoietin, thereby becoming a part of one of the biggest battles in biotech, the fight between Amgen and Genetics Institute for ownership of the erythropoietin patent.)

Whilst at Caltech, Mike had been heavily involved in the Hood and Dreyer labs' attempts to develop a protein sequencer that overcame the problems inherent in the design of Pehr Edman's original machine. Despite incremental improvements to the spinning cup sequencer that speeded up the reactions, nobody had been able to do anything about the big problem, the amount of pure material that was needed to get sequence. There was also a secondary issue, that short peptides could not be sequenced because they tended to disappear along with the reagents when the samples were washed. Because the best method for sequencing a large protein was to cut it into smaller peptide fragments and use these fragments as the starting material for sequencing, this was also a major drawback.

The solution to both issues, which Mike had been working on at Caltech and which was further elaborated after his departure, was to change the way the protein substrate was presented to the chemicals of the Edman degradation. In Edman's sequencer, the protein was spread on the inner surface of the spinning cup and held there by centrifugal force, but if instead it was adsorbed onto a glass fibre disk in a small reaction chamber, losses during the reactions and the washing steps were greatly reduced. In the Caltech machine, the key reagents for the cleavage reaction were delivered to the protein substrate in streams of gases, and washing and extraction of the cleaved amino acid was done by adding judicious amounts of liquid organic solvents, all designed to minimise sample loss.

When combined with advances in identification of the amino acids coming out of the sequencer, the sensitivity of the new gas-phase machines was astonishing. Model 470A, the first commercial gas-phase sequencer sold in 1981 by Applied Biosystems, the company cofounded by Hood and Dreyer to exploit their work, required a thousand times less sample than the spinning cup models and had a comparable yield. Suddenly, for the tiny number of labs in the world in possession of a gas-phase sequencer, it was possible to think about sequencing an entire new world of lower-abundance proteins. Mike, sitting at the ICRF with his own version of the Caltech machine, together with a heavily pimped commercial Beckman spinning cup sequencer, was approached by two labs wishing to solve a really big, really interesting problem: unmasking the hidden secrets of growth factors and their receptors.

In the years since the discovery of NGF and EGF, work on growth factors had spread beyond the confines of embryology labs into another field, that of Rita Levi-Montalcini's friend Renato Dulbecco. The labs studying the properties of normal and transformed cells in culture knew that in addition to the chemicals found in cell growth medium, tissue culture cells could only thrive in the presence of serum, the liquid portion of clotted blood. Serum, besides containing nutrients, was also a source of other substances that seemed to be involved in instructing cells to divide and, sometimes, to change their shape and function; in other words, it contained growth factors.

After jogging along together in possession of their shared knowledge for a while, there was a collision between the growth factor and tumour virus fields in the late 1970s. It had been known for a while that transformed cells growing in culture had less need for serum than their normal counterparts, and in 1978, Joe DeLarco and George Todaro at NIH showed why; cells transformed with RNA tumour viruses were releasing their own growth factors, rather than being dependent on those found in serum. Furthermore, the transforming growth factors were able to function through the cell surface receptors for normal growth factors and could stimulate the growth of normal cells. This finding, that growth factors might mediate the action of oncogenes during transformation, led to Mike Sporn and George Todaro's autocrine hypothesis of 1980, which proposed that cancer cells could escape normal growth controls by producing and using the growth factors required for cells to proliferate. Further data from many labs showing that the tumour growth factors could elicit many of the same reactions within cells as did normal growth factors gave rapid validity to Sporn and Todaro's idea.

One of the normal growth factors that featured heavily in the flurry of papers surrounding the publication of the autocrine hypothesis was Platelet-Derived Growth Factor (PDGF). PDGF had been discovered in 1971 by Sam Balk at Rockefeller University in New York. Whilst studying the growth of transformed and normal fibroblasts, Balk, and a number of long-suffering cockerels who were bled weekly for his experiments, showed that serum, rather than plasma, the liquid made from unclotted blood, contained the extra ingredient required to keep normal fibroblasts happy. Three years later, two other labs confirmed Balk's initial observations and added the vital detail that the growth factor was being released from platelets during the clotting process. The newly christened PDGF attracted a lot of attention, because it seemed likely to be important, in a good way for wound healing, and in a bad way for atherosclerosis, two processes in which platelet aggregation and clotting were intimately involved. It also seemed like a great candidate to try to purify, because human blood and its products, including platelets, were widely available from blood donor banks.

Concerted efforts to purify PDGF started in 1976, and in 1979, Chuck Stiles, Chick Scher, and Harry Antoniades in Boston, and Calle Heldin, Bengt Westermark, and Åke Wasteson in Uppsala published two independent papers reporting their success. In the next few years, the Uppsala PDGF lab went down the biochemical pathway, showing that there were receptors for PDGF on a limited spectrum of cells derived from connective tissue, and that adding PDGF to these cells resulted in an increase in phosphorylation of cellular proteins, one of the intracellular effects that both normal and transforming growth factors were able to elicit. The Boston grouping of Stiles, Scher, and Antoniades was more interested in biological data, demonstrating amongst other things that if cells were infected with tumour viruses, the requirement for PDGF was lost. They also started looking at whether PDGF could affect the transcription of cellular genes, reasoning that any lasting changes caused by growth factors had to occur through synthesis of new proteins. Other labs joined in the fun, including that of Tom Deuel, a clinician scientist working at Rita Levi-Montalcini's old stamping ground, Washington University in St. Louis, Missouri.

PDGF was also being studied at the ICRF in London, although in a slightly different form. To complement his own interest in growth factor control, Michael Stoker had hired Henry Rozengurt, whose lab was working on cells that made a PDGF equivalent called Fibroblast-Derived Growth Factor (FDGF). The Rozengurt and Waterfield labs had come together with the aim of purifying and sequencing FDGF, but even taking into account the reduced appetites of Mike's modified sequencers, the protein was just not made in sufficient quantities for the project to be feasible, and it was abandoned. However, the FDGF project, together with prolonged exposure to the work going on in Stoker's lab, had turned Mike on to growth factor research, and he decided to take on PDGF himself. As a better source of material, the lab started purifying PDGF from pig platelets, but in 1981, Mike got a much better offer. He was approached independently by Tom Deuel and Calle Heldin, who both wanted to set up a collaboration in which they would provide human platelet PDGF for Mike to sequence. Thinking that two sources of protein would be better than one, and also being someone who didn't like to say no, Mike agreed to both proposals. Deuel and Heldin, previously competitors, found themselves in a three-way collaboration, which, rather surprisingly, worked.

The Deuel and Heldin labs, soon joined in their efforts by Paul Stroobant, Mike's South African postdoc, set about purifying enough PDGF to analyse and sequence. The Swedes were fortunate in that they could get platelets from healthy donors, courtesy of the Finnish Red Cross, but the London and St. Louis labs were using hundreds of litres of outdated platelets from U.S. blood banks, which were at that time pretty dubious sources. The U.S. dependence on paid blood donation meant that their blood banks were top heavy with donations from the desperately poor, who were more likely to be ill with blood-borne diseases such as hepatitis, and a sinister new illness, Acquired Immune Deficiency Syndrome. As Mike recalls: "Little did we realise that HIV was beginning. At that time, safety measures were not good at all—I mean, we did dress up, but there was no way that we weren't getting platelet and serum-derived products spread about in a way that nowadays you wouldn't even begin to do." Fortunately, no one seems to have suffered any ill effects.

Getting PDGF into a suitable state to run on the ICRF sequencer was a knotty problem. During the purification process, proteins can start to degrade, attacked by enzymes called proteases or simply falling apart of their own accord. Sequencewise, this is a disaster, because the Edman degradation relies on the protein substrate having a completely perfect front end, or amino terminus. If a protein has a frayed amino terminus, what comes out of the machine is a hopeless mix of amino acids, rather than a single one, making identification impossible. The first purification attempts of the three collaborating labs produced a PDGF prep containing five different amino termini, clearly a mix of five different polypeptides. It was hard to tell whether PDGF was some kind of protein complex, with different polypeptide chains, or a single protein that fell apart very easily; but whichever it was, the polypeptides had to be separated from each other before they could be used as sequencing fodder.

At the same time as his collaborators and lab members were wading around in litres of dodgy blood products, Mike had been thinking about another issue that would become of great importance once PDGF was purified: how were they to draw any conclusions about PDGF's function from its amino acid sequence? Seeing if it resembled other proteins of known structure and/or function was the best bet, but such information was hard to come by. Bioinformatics, the mining of data from sequences and other biological information, hardly existed; the sheer dearth of sequenced proteins meant that the only database of the 1970s, the Atlas of Protein Sequence and Structure, proprietor Margaret Dayhoff, only contained 310 unique sequences. Mike had done a bit of work already on database development and searching during a sabbatical at the Harvard Biolabs (taken to accompany Sally, who was doing a medical residency at Massachusetts General Hospital) so he was well aware of its importance. Therefore, in 1983, he hired the New Zealander Peter Stockwell as the lab's resident bioinformaticist. Stockwell's remit was to develop programmes for creating and managing sequence libraries and to help with database searching using the state-of-the-art DEC 20/50 mainframe, a magnificent beast with a computing capacity several times smaller than that of a mobile phone. He was assisted in his labours by Mike's technician Geoff Scrace, one of whose jobs was to go to the library

Mike Waterfield at the computer, ca. 1985.
(*Photograph courtesy of CRUK London Research Institute Archives.*)

every week, find all of the new and interesting sequences, and get them into the database.

The Waterfield laboratory protein sequence database was not the Dayhoff Atlas of Protein Sequences, but a far superior product, NEWAT, first described in 1981 by Russ Doolittle, a protein chemist working in San Diego. Doolittle's main interest was in molecular evolution, comparing protein structures across different species, and he had gotten so tired of the Dayhoff database's limitations that he decided to make his own. Because he was in charge of childcare most afternoons after school, he roped in his 11-year-old son Will to do the typing, with his secretary Karen helping with proofreading. Together, the three of them scoured the literature for the sequences of every interesting protein and put together a simple search algorithm so that new proteins could be compared with existing ones. As word spread, Doolittle received hundreds of requests to run protein sequences against NEWAT. In the beginning, he ran the searches himself and then notified the interested parties if there were any matches, but eventually, searching was taking up so much time that he began to give out copies of NEWAT on tape. Not surprisingly, he was a little conflicted about parting with his precious database: "Certainly the first few times, I gave it a lot of thought, because we had gone to a lot of trouble to put this together and now anybody could do what I could do—it was like giving away the golden wand. On the other hand, scientists are very sensitive about their image, and I would never want anyone to think I was the sort who wouldn't share. I also realised that they weren't likely to keep it up to date so I'd always have a little advantage over people. So I did it willingly and after a while it just became routine" (BBC 1983).

On 4 May 1983, Mike phoned the Doolittle lab and got through to Karen, Doolittle's secretary, who, once Doolittle had cleared it, sent a tape of the latest version of NEWAT to London. It was eagerly awaited because the PDGF project had arrived at the point where the protein was pure enough to be entrusted to Nick Totty and Geoff Scrace, the guardians of the sequencers. Two years on from the start of the collaboration, the lab was finally starting to get some very good data.

The tape arrived and was handed over to Peter Stockwell, who loaded up the new database, typed in the PDGF sequence, and set the sequence comparison programme going. What came out was truly remarkable. Human PDGF was an almost perfect match to $p28^{sis}$, the product of the v-*sis* oncogene of simian sarcoma virus, and the growth factor and cancer fields had finally, irrevocably fused.

The realisation that $p28^{sis}$ was PDGF was the keystone completing the bridge of data that led to the understanding of how growth factors contributed to cancer. Once it was discovered, it seemed blindingly obvious that, rather than going to the trouble of evolving a new alien growth factor, an RNA tumour virus could just steal a normal cellular growth factor and use it to force cells to proliferate uncontrollably. There was ample precedent for the retroviral theft of cellular genes, but nobody before had figured out the true function of any of the unwilling captives.

Mike recalls vividly what happened: "When Peter found it he came in from the computer and showed me, and we were tremendously excited, because one just doesn't find this sort of observation unless it is something extremely significant; the degree of identity we found was very unusual. I ran out to talk to a couple of my friends who were virologists, to show them and confirm my feelings about the significance of this observation, and of course they immediately said yes, this was something that was extremely important and interesting."

Amidst all the excitement, Mike also realised that he and his colleagues would really need to get moving: "The statement '$p28^{sis}$ is PDGF' meant everything to almost every molecular biologist involved in the business. It was like a newsflash, so should we tell everybody, or what should we do? We decided we should write the paper describing our observations and at the same time try and complete the experimental work supporting our observations."

Getting scooped under such circumstances is the gnawing fear of all scientists, and in the 27th May issue of the journal *Science* (confusingly published on 20 May), Mike saw with sinking heart an article entitled "Human Platelet-Derived Growth Factor: Amino-Terminal Amino Acid Sequence," whose authors were Harry Antoniades, one of the Boston PDGF gang, and Mike Hunkapiller, Lee Hood's right-hand man at Caltech. However, feverish scanning of the paper revealed the unbelievable—Antoniades and Hunkapiller had completely missed the $p28^{sis}$ match. The game was still on, but only just; with both sequences out in the public domain, someone else was bound to find the match in very short order. Mike called Peter Newmark, the biology editor of *Nature*, *Science*'s great rival, to ask if he was interested in rapidly publishing a paper that was not only very important but would make *Science* look a bit silly. Newmark, not surprisingly, said that he was. Mike sat down and wrote a draft of the paper at his dining table in Gospel Oak and then did something that in retrospect was extremely brave, or possibly foolhardy: He and Sally went on holiday to Rhodes, to the little village of Lindos, famed for its ancient acropolis, but not big on telephonic access to the outside world.

At about the time that Mike and Sally were taking their first stroll around Lindos, Russ Doolittle in San Diego was reading Antoniades and Hunkapiller's paper as some light Saturday morning home entertainment. Being who he was, he too ran a database search with the PDGF sequence and got the same startling result that the Waterfield lab had seen a fortnight earlier:

> It was extraordinary. It was a magic moment for me because not only did I find something that matched, but I knew the match. I was familiar enough with this area of biology that the impact was tremendous. It was very interesting, and I'll never have another moment like that, I'm sure. I took several deep breaths and I had a number of different thoughts. I must confess, among them was "I wonder if anyone else knows this?"—I probably hoped that no one did. And I reflected on who might know. The article had said in it they had conducted a limited search. It wasn't quite certain what that meant, but I thought "well, if somebody else knows, they know, but in the meantime, I'll certainly tell people what it is" (BBC 1983).

As was his standard procedure when he found NEWAT matches, Doolittle wrote to Stu Aaronson, whose lab had reported the v-*sis* sequence, and Hunkapiller and Antoniades. The two groups quickly got in touch, and after some frantic work to extend the PDGF amino-terminal sequence, started preparing a very short paper to be submitted to *Science*. However, the new paper had acquired several extra authors, one of whom, Lee Hood, was not noted for his ability to keep scientific secrets. Knowing that he was due to give a talk at a big conference in San Francisco on 3 June, he called up Russ Doolittle and asked him if it was OK to mention the PDGF–p28^{sis} homology. Doolittle reacted with commendable mildness, because it was his find, and really, he should have been the one to reveal it to the world:

> I wasn't sure if I minded or not to tell the truth, but I said, "if it's fine with everybody else...." I was concerned, I must tell you, as we hadn't written anything down about this yet, and without having any idea there were any competitors, it just seemed to me indiscreet to be mentioning it before you had published anything or at least had submitted something. At any rate I suspected all the others said the same thing, that it was OK with everybody else, and he mentioned it, to a packed and overflowing crowd I hear (BBC 1983).

Sitting in the second row of the conference hall for Hood's talk was Tom Deuel, Mike's collaborator. It must have been an awful moment for him, although his clinician's ability to dissemble comes to the rescue: "My feelings were very mixed. I was surprised, I was disappointed, I was excited. It was very nice, in the sense of seeing our finding confirmed by others ... [but] we were of course disappointed. We wanted to be the first ones to have it presented to the public" (BBC 1983).

Once out of the session, Tom phoned the Waterfield lab and discovered to his horror that Mike was off on holiday. Once he'd told the lab what had happened, they dispatched a telegram to Mike, simply stating "PLEASE CALL LAB." From amongst the feta cheese and olives in the local supermarket in Lindos, Mike did just that, and told Paul Stroobant to get the paper off to Peter Newmark at *Nature* as quickly as possible. The next day, another telegram arrived, this one simply announcing "NEWMARK READY PROBABLY MONDAY PM. FOUR WEEKS OUT."

The American paper, with Russ Doolittle as first author and a single figure showing the match between PDGF and p28^{sis}, was submitted to *Science* on 3 June, and accepted on 12 June. Mike's lab, still in his absence, got their paper, a rather longer affair, to *Nature* on 14 June,

ΟΡΓΑΝΙΣΜΟΣ ΤΗΛΕΠΙΚΟΙΝΩΝΙΩΝ ΤΗΣ ΕΛΛΑΔΟΣ Α. Ε.

OTE

ΤΗΛΕΓΡΑΦΗΜΑ

Ἐνδείξεις Λήψεως	Γραφεῖον Καταγωγῆς	Ἀριθμός	Λέξεις	Ἡμερομηνία	Ὥρα	Ἐνδείξεις Μεταβιβάσεως
	LONDON	A14261	10	9/6	16.41	
08.40	DR. M WATERFIELD STUDIO VALENTINO LINDOS RHODES					

PLEASE CALL LAB

K. A. 815.58.13 - 12/75

Holiday disruption: The telegram Mike received from his laboratory. (*Courtesy of Mike Waterfield.*)

where it was accepted three days later. The speed of both journals was completely unprecedented, and the papers were eventually published by *Science* on 15 July (although, as mentioned above, this really meant 8 July), and by *Nature* on 7 July, a testament to how intense the competition between the two journals was at the time. However, by the time of the papers' publications, both camps were enveloped in a media storm, thanks to an indiscreet and overeager journalist who had heard Hood's talk and written it up for the 18 June issue of a magazine, *Science News*.

Shortly after *Science News* broke the story, Russ Doolittle received a phone call: "A reporter from the *Los Angeles Times* called up here, and said 'See here, I see your name mentioned in this thing. Would you like to tell me about your role in it?' Well ... our public relations people here said 'Fine, talk to reporters. It's too late now—there's no embargo any longer.' " Doolittle's press contact also told him about the rival paper:

> He said, "By the way, I just talked to someone at the Salk Institute and they told me there are some people in England that have discovered the same thing." And he told me the name of the person, which I confess I didn't recognise at the time. And, well, I guess ... there was some disappointment to find that somebody else was on top of it so quickly. On the other

> hand, it happens, and that's the way it is.... It wasn't until later that evening that it dawned on me that the name that had been told to me was familiar, and I went through my records and found that Dr Waterfield had requested one of the tapes. It was ironic ... I guess I was relieved that our database had played some role in their discovery also, and that sort of made it nice again I guess.

Would he have liked Mike to have informed him of the homology? "I've seen it both ways, and yes, I would like to have known, but I also quite understand why he didn't tell anybody.... Given the nature of the work they were doing, it seems to me that most people would have behaved the same way. The one thing I think I wouldn't have done if I were he, is that I would never have gone off on holiday until I had finished the task!" (BBC 1983).

Once the story was out in the *Los Angeles Times*, it was picked up by the rest of the U.S. media, and there was no holding it. Mike, finally back from his holidays, got a call: "We were asked by the BBC World Service if we had done this particular piece of work because they had seen it in the *Herald Tribune*.... I think we would much much rather have held a well-ordered press conference at which we could fully explain everything, but we did the best we could."

The ICRF, with a nonexistent press office and limited experience in dealing with the media, put out a press release that Peter Newmark subsequently described as "not cautious. It did not mention the fact that the only strong link of the discovery to cancer was in the case of tumours of monkeys caused by simian sarcoma virus. It claimed categorically that PDGF 'produced by the oncogene in excessive amounts ... leads to uncontrolled growth in ... bones and other connective tissue' and that drugs can now be devised to block the effects of PDGF and 'perhaps stop cancerous growth'" (Newmark 1983).

Mike, without the protection of a public relations department, found himself besieged by a press pack that had scented a miraculous cure for cancer. At work, it wasn't too bad: "It got picked up when there wasn't a lot of news. ICRF was invaded by a lot of newspapers including some journalists from Brazil who were there to cover a football competition. We had these guys come in the lab and they were filmed. We dressed them up in lab coats but they were football correspondents!" At home, however, it was a lot more alarming. Mike and Sally were doorstepped aggressively by reporters: "It was quite upsetting because I was very naïve and open about it and then got burnt—Sal and I ended up with our picture on the front of the *News of the World* which was taken because the guy was banging on our door in Oak Village, and we let him take the picture in order to make him go away."

Although many of the papers managed to produce a reasonably accurate account of the work following the press release, others did not. Mike and Sally's nemesis, *The News of the World*, was especially irresponsible, leading with the headline: "Cancer Jab on the Way," dubbing poor Mike a "superboffin" and quoting him as saying that "within the year, or at least by this time next year, we will find a drug to halt some kinds of cancer. It would be nice to say a quick pill will solve the problems, but I don't think that's so.... I think it will have to be an injection." The story ended with the tale of a little girl, Victoria Hart, who had just died of bone cancer, and implied that her cancer might have been cured by the mythical cancer jab. It was a monument to the reporting style that led to the paper's eventual closure in 2011.

SUPERBOFFIN Dr Mike Waterfield and his wife Sally

CANCER JAB ON THE WAY

A SUPERJAB that will cure certain types of cancer will be available in Britain within a year.

The amazing breakthrough was revealed yesterday by pioneering British scientist Dr Mike Waterfield.

News of the World Reporter

And he pledged that despite the appalling lack of Government money available for cancer research in this country, we would still win the day . . . with a cure for cancer.

Dr Waterfield, of the London - based Imperial Cancer Research Fund, expects an overall cure within five years.

But in the meantime, certain kinds of cancer which attack bone and tissue could be solved much more quickly with the new injection.

Hopes

Dr Waterfield, 42, of Hampstead, London, hit the headlines yesterday with the amazing discovery which he made with an international research team.

He said last night: "While there's no way we want to give false hopes to people, I would honestly hope that within the year, or at least by this time next year, we will find a drug to halt some kinds of cancer.

"It would be nice to say a quick pill will solve the problems, but I don't think that's so. However, I can say that I think we'll get together a drug that can halt the growth of cancer cells.

"I think it will have to be an injection.

"But it would be improper of me to stand up and say to anyone suffering from cancer today that we can help them. But that five years is going to make a lot of difference."

Dr Waterfield's team found that a substance in the blood which aids the recovery of wounds can go "rogue" and cause cancer.

He said yesterday: "I can't take credit for any major breakthrough. It's been a great deal of teamwork and I regard this as and I regard this as another link in the chain."

The breakthrough would spell hope for tragic children like Victoria Hart, who fought a lifelong battle against a rare bone cancer.

Victoria, 12, of Billingshurst, Sussex, forgot her own suffering when Household Cavalry horse Sefton was injured in the Hyde Park bomb attack last year. She sent a get well card.

On Friday Victoria was buried, and among the flowers at the funeral was a horseshoe from Sefton.

Dr Waterfield said the cancer that killed Victoria was among the types his team were dealing with.

WEATHER

SUNNY and warm. Max temp 24C (75F). Outlook: Fine.

Article from *News of the World*, 3 July 1983, © NI Syndication.

Mike emerged bruised and blinking from his ordeal by media, and life returned to some semblance of normality by the autumn of 1983. Work on PDGF continued, but other things were also quietly afoot in the Waterfield lab. In line with his long-standing interest in membrane proteins, acquired in his days at Caltech working on immunoglobulins, Mike had been running another project in parallel with the PDGF work: to purify and sequence the receptor for Epidermal Growth Factor, the opener of baby mice's eyes.

Nearly 20 years on from Stan Cohen's original isolation of EGF, addition of the protein to cells was known to induce the export and import of many substances, stimulate cellular metabolism and proliferation, cause changes in cell shape and behaviour, and activate DNA, RNA, and protein synthesis. How EGF did any of these things was more or less a mystery, but binding of EGF to its cellular receptor was an absolute requirement. The receptor was an intriguing molecule. It had a tripartite structure, with an extracellular EGF-binding domain, a transmembrane region, and an internal domain, which the Cohen lab showed in 1980 possessed tyrosine kinase activity, the ability to identify the amino acid tyrosine in a protein and stick phosphate groups on it. The receptor's tyrosine kinase activity was tightly regulated—EGF binding was the "on" switch, whereas in the absence of EGF, activity was firmly off.

The finding that the EGF receptor had tyrosine kinase activity was the first biochemical link between oncogenes and growth factors, giving substance to the theories suggesting that both might work to stimulate cell growth through similar pathways. Protein kinase work had become obligatory in the oncogene world after the laboratories of Mike Bishop, Harold Varmus, and Ray Erikson discovered in 1978 that v-Src, the oncoprotein of Rous sarcoma virus, had protein kinase activity. Phosphorylation of proteins can alter their functionality dramatically, and the idea that oncogenes might be able to tamper with such a fundamental control mechanism was immensely attractive. Accordingly, everyone looked to see if their pet protein could join the kinase club. It was on such a quest that tyrosine kinases were discovered by Tony Hunter in San Diego, during the three-way race to define the function of the middle T oncogene of polyomavirus. Hunter and Bart Sefton went on to show that v-Src and its cellular progenitor c-Src were both tyrosine kinases, and using their assay, tyrosine kinases started popping out of the woodwork in all directions, including in the growth factor world. The EGF receptor was the first so-called Receptor Tyrosine Kinase to be categorised, but was soon joined in 1982 by the receptors for insulin and PDGF, suggesting the existence of a common pathway for intracellular signaling by some growth factors. The Waterfield lab's discovery that PDGF was the $p28^{sis}$ oncoprotein was the punchline for the whole story to date, showing how this normal pathway could be subverted.

Although the EGF receptor had an extremely trendy function ascribed to it, it had yet to be sequenced, and taking on such a challenge was exactly what Mike and his lab loved. Working in their favour was the fact that they could tap an unexpectedly rich source of protein; A431 cells, a human squamous carcinoma line, had been shown by Joe DeLarco to carry millions of EGF receptor molecules, as opposed to the normal number of a few thousand. This had not been overlooked by the rest of the field, and the Cohen lab had begun purifying the EGF receptor in the late 1970s by running A431 membrane extracts down an affinity column to which EGF protein had previously been chemically coupled. Under the appropriate conditions, the EGF proteins could hook their receptors out of the extracts as they passed through the column, whilst everything else simply flowed past. The EGF receptor molecules could then be washed off the column by changing the binding conditions. However, even the great Stan Cohen had been unable to get any sequence by this method because the yield was too poor. At the ICRF, Mike, looking around for a better alternative, got talking to his next-door neighbour, Peter Goodfellow.

Peter Goodfellow's main interest was in human genetics, as will be related in Chapter 8, but in the early 1980s, freshly arrived at the ICRF, he was much in demand for another of his skills—making monoclonal antibodies. As shown to great effect for SV40 large T and p53, monoclonal antibodies were and are by far the best tools for pulling specific proteins out of complex mixes, and after having a long discussion with Mike, Peter came up with a very cute way of making a monoclonal antibody against the EGF receptor. Rather than dithering about making huge quantities of purified receptor and injecting it into mice in order to get an immune response, why not simply immunise mice with A431 cells? There was a very good chance that that would work just as well, given that the cells were fairly bristling with receptors, and it would be an awful lot easier and faster. Monoclonal antibody–producing cells (hybridomas) could then be made, and the antibodies they secreted tested for their ability to block EGF binding to A431 cells; if the monoclonal was specific for the EGF receptor, it would sit on it, thereby hiding it from EGF and giving a negative result in an EGF-binding assay. The trick worked rather well, and a joint paper describing EGFR1, the first specific anti-EGF receptor monoclonal antibody, was published in September 1982, with the Waterfield lab, in particular Mike's postdoc Brad Ozanne, doing the binding assays, and George Banting from Peter's lab covering the antibody work.

In October 1982, Mike acquired a new graduate student, hired to join the EGF receptor sequencing project. Julian Downward arrived at the ICRF trailing a veritable wagonload of British establishment baggage; he'd been to Eton and Cambridge, and his Dad was a Major General in the British Army. Fortunately, none of these things managed to obscure

Julian Downward, 1984.
(*Photograph courtesy of Peter Parker.*)

the fact that Julian was both extremely nice and so self-effacing that it was easy not to notice just how good a scientist he was. He'd come to the ICRF on the recommendation of his lecturers, who'd characterised the place as a land of milk and honey compared with their impecunious laboratories in the windswept wastes of the Fens, but when he arrived Julian found that things weren't much different to the Cambridge Biochemistry department:

> I thought that everything would be shiny and people would be walking around in designer outfits or something, but it was actually fairly similar to Cambridge. In Mike's lab, I remember that a lot of people had beards, so it had a sort of chemistry feel to it. He had these two senior technicians, Geoff Scrace and Nick Totty, who were very beardy, and their job was to run all these sequencers. It did seem quite impressive—the machinery was all very impressive. I got a tremendous sense of enthusiasm from Mike, which was very endearing. He said cancer was bound to be a subversion of the way normal growth mechanisms worked and this was the way to find out.

"Very beardy." Nick Totty and Geoff Scrace, ca. 1984.
(*Photograph courtesy of Peter Parker.*)

Mike realised quite fast that he'd lucked out with his new student:

> He came in on day one and decided on the project and said, "Right, I want to get started." I said, "What help do you need?" and he said, "None, I'm just going to get on with it." And he quietly worked his butt off! I think [he was] someone who was just focussed, fascinated and extraordinarily capable—he obviously had a CV that said he'd done rather well at Eton and rather well at Cambridge—I think he'd passed out first in both places. Remarkable guy, and he was also quite a funny mixture. Now, I don't regard him as being socially wild ... he's cool and calm. But in those days during his PhD he was really wild at times! We used to go down to conferences at Wye College, and I had some amazing times with Julian. Over the years he's settled down a bit—maybe too quickly!

The EGF receptor project had been going for a little while, with a post-doc, Elaine Mayes, doing protein stuff using A431 cells, and a PhD student, Nigel Whittle, trying to clone the EGF receptor gene directly. Julian's job was to work with Elaine on receptor purification, but instead of using A431 cells, Mike wanted him to try a different source. Some years back, it had been noticed that there were lots of EGF receptors on placental membranes, and Sally knew a gynaecologist at Queen Charlotte's Hospital. Putting these two facts together led to the 22-year-old Julian being sentenced to collect fresh afterbirths from the new mothers of Hammersmith, an experience he remembers vividly after more than quarter of a century: "Getting it from placenta was just a complete nightmare. I used to go over and hang around all day waiting for these things to come. I'd sometimes do a few at a go depending on what was available. I would be waiting in the sluice room with loads of ice. It wasn't very nice at all!"

Once the placenta had been transported back on the tube, Julian would settle in for a long evening in the lab:

> Placenta's a bit looser than liver but it's that sort of thing. The first step in the process was you'd cut off the gross membranes and then with all the meaty bits you'd cut them into little chunks in cold PBS [phosphate-buffered saline] and put them into a blender. Then you'd spin that stuff down and put it into a low osmotic strength buffer for an hour. The idea was that these villi would bleb off and so you'd get this preparation of villi, and those were very rich in EGF receptor. So you would end up with a nice clean membrane fraction from the surface of the syncytiotrophoblast. Once you got those membrane preps you could then store them and freeze them. Usually on those days if I could get away from the hospital about four in the afternoon I would be finished by midnight, but if it got much

> later than that it would just go on and on. And there was loads of mercaptoethanol and everything else and it stank.

After a few months of dicing and slicing, Julian was getting small amounts of pure placental EGF receptor and had also discovered that Peter Goodfellow's EGFR1 monoclonal antibody was a far better purification reagent than were the old Cohen-style EGF-binding columns. However, in the absence of a West London population explosion, Julian had also realised that it was going to be very hard to get enough placenta to make sufficient EGF receptor to sequence. The A431 cells were clearly the way to go.

Putting placenta on the back burner was surprisingly easy, thanks to the PDGF and $p28^{sis}$ story going nuclear about six months after Julian's arrival. Many of the people in the lab, including Elaine and Nigel, who were both tired of fighting a losing battle with the EGF receptor, dumped their projects in favour of PDGF-ology, and Julian was left to carry on by himself: "I did think about jumping too, but then I thought 'Well, actually this is quite an interesting project in its own right, and it would make life a lot simpler if it was just open to me.' So I started doing a lot more on the A431 cells. I started producing huge amounts of cells with the cell culture people here—I started getting them to do eighty 2.5-litre roller bottles a week. The cells grew like weeds and you got vast amounts of cells and they had 2 million receptors per cell—the amount you got was just unbelievable. Suddenly it all started to come together pretty well."

Despite Julian's having to deal with a bathtub's worth of cells every week by himself, getting lots of EGF receptor was in the end quite straightforward, thanks to the EGFR1 magic monoclonal antibody. He even managed to get something from the placental preps, which he kept doing, "because I just wanted to feel that the placenta had led somewhere." The only real problem was the cold room, where Julian spent a large part of his time mollycoddling his protein. Keeping warm whilst doing experiments in there meant having to dress as though it were snowing, even in high summer, but as an additional hazard, the fourth floor cold room was also liberally anointed with fresh cow brains. Peter Parker, a recently arrived postdoc with Mike, was trying to purify an enzyme called protein kinase C from the brains, so fellow users had to breathe a pungent aerosol of brain and mercaptoethanol, singeing their sense of smell and potentially exposing themselves to mad cow disease all in one go.

By the end of the summer of 1983, as Brazilian football reporters, assorted gutter journalists, and BBC film crews surged through the laboratories in pursuit of his superboffin boss, Julian had got enough protein together to put onto the sequencer, under the watchful eye of Nick Totty. He had also acquired a temporary collaborator from the Weizmann Institute in Israel, a PhD student called Yossi Yarden. Mike had set up a collaboration with Yossi Schlessinger, Yossi Yarden's boss, thinking that he might need another source of EGF receptor should Julian not get much from A431 cells. The Israeli source was superfluous to requirements by the time Yossi arrived, but the few weeks of additional company weren't unpleasant:

> Yossi is very taciturn and he's got this big hearing problem from when he got blown up while he was in the Israeli paratrooper brigade doing his national service. He was a nice guy, very straightforward. When they finally got their stuff onto the sequencer it didn't really give anything that was obviously the EGF receptor. What we did get out of it was ubiquitin, which at the time we just thought was some contaminant. Actually, it probably was interesting, as the receptor is ubiquitinated, and now there's a whole industry about ubiquitination of the EGF receptor.

Yossi eventually returned to Israel without getting anything sensible in the way of sequence, but went down in lab mythology for his ability to fall asleep instantly and anywhere, as Mike remembers: "He used to occasionally spend the night sleeping on the bench in the lab because when he'd been in the Israeli army he used to run miles and miles with a pack and rifle across the desert—he always said he could sleep even when he was running. He was a wonderful guy—quiet, determined."

By the winter of 1983, Julian had around 250 amino acids-worth of sequence from the EGF receptor, but the protein had had to be chopped into smaller peptides for sequencing, and there was no way of knowing how the resulting short sequences fitted together; the words were all there, but the story they told could not be deduced. This wouldn't matter if the peptide sequences matched something in the Doolittle database, but disappointingly, when Peter Stockwell ran a search, there were no homologies to anything. It looked as though Julian would be in for the long haul of figuring out what possible DNA sequences coded for the peptides and using these to go on a fishing expedition for the EGF receptor gene in a human DNA library. The methodology for all this was still very new, and so Mike roped in a friend of his to help. Axel Ullrich from Genentech, the company that Herb Boyer and Bob Swanson had set up to exploit recombinant DNA technology, already worked on EGF, was a cloning expert, and was very keen to get on board. However, before Ullrich could get started on the project, Julian hit pay dirt:

> A paper had come out from a guy in Japan [Tadashi Yamamoto and colleagues] who had been sequencing the v-*erb-B* oncogene from AEV [avian erythroblastosis virus]. So Geoff Scrace put the sequence into the database. Every time Geoff updated the database I'd just do a sequence search without relying on Peter Stockwell. I used to do it late at night because that's when Peter went home and I could get onto the computer terminal (there was only one computer terminal!). If I did it when he was around he was really grumpy and he used to shout at me because I was disturbing what he was doing. So I would wait till he went home.
>
> I just did it one evening—every couple of weeks I'd check—and then suddenly saw these things all lined up. You know, the first one, I thought maybe that's chance, but just more and more kept coming up on the screen. There were about fifteen to twenty sequences and about half of them were lining up. They all just started popping out and it was very obviously a big deal—the matches were perfect—I mean not totally perfect but every peptide aligned—90% or so homologous. The v-*erb-B* oncogene is truncated—it's lost most of the amino terminal part but it's got the transmembrane region and most of the intracellular bit. So it was about half of the EGF receptor represented there.
>
> So then I called Mike at home—it was probably about nine in the evening in early December. Mike realised very rapidly this was something important so he came in with Sally and we spent the next few hours tinkering around with these sequences, and then he drove me home.

Discovering something quite so important after only 15 months as a PhD student is something most scientists can only dream of, and Julian was well aware of the significance: "At the slow times in the work before that I had thought how difficult it was to achieve anything—I remember sitting up in the library reading copies of *Nature* and thinking 'there's *so* much work here—how can anybody achieve all this in a PhD?' But then when it all happened I knew it was a big deal. And then Paul Stroobant said 'This is something you should savour because probably you'll never do anything as important as that again.' It was pretty reasonable advice. I appreciated how important it was."

Julian and Mike spent Christmas 1983 writing up the work, and submitted a paper entitled "Close Similarity of Epidermal Growth Factor

Mike Waterfield and Yossi Schlessinger, ca. 1985.
(*Photograph courtesy of Martin McMahon.*)

Receptor and v-*erb-B* Oncogene Protein Sequences" to Peter Newmark at *Nature* in mid-January 1984. Perhaps fearing a repeat of the PDGF saga, Newmark accepted the paper in five days and published it a fortnight later, but this time, there were no nasty scientific surprises in store.

Having been thoroughly shaken by the attentions of the press the previous year, Mike made sure that he wasn't in London when the paper came out: "I waltzed off to Israel and gave two big lectures at Tel Aviv university, and had a great time!" Sally was left behind to fend off the journalists by herself, because Julian was also absent. Mike had delegated a rather important task to him:

> Mike got invited to a UCLA meeting in Steamboat Springs [a ski resort in Colorado] called something like "Genes and Cancer," and for some reason he didn't want to go. So he said (looking back on it, it's fairly unbelievable!), "I've got a 23 year old graduate student, he'll be perfect to give a talk at this major meeting." It was the first talk ever that I'd given. We put together a presentation and I actually got a lot of advice from Mike Fried. Mike decided that Mike Waterfield was not taking this sufficiently seriously and I would be eaten by the lions unless I was highly trained up to give this talk. He was very helpful and gave a lot of advice going over this talk and how to present it.
>
> So I headed off to this meeting, which was shortly before the paper came out in the middle of February, and gave this talk, which made a big splash. Tony Hunter was an organiser of the meeting and I remember he invited me to dinner with him and some other people in this fancy restaurant in

the resort and I thought this was amazing because I knew Tony Hunter's work from my undergraduate days, and subsequently.

Tony Hunter, the kinase king, remembers Julian's talk at the Steamboat meeting very well. He'd heard on a visit to London in December 1983 that the Waterfield lab had a lot of EGF receptor sequence but the homology with the v-*erb-B* protein had not yet been noticed. When Julian got up to speak, his training in the Mike Fried Seminar Bootcamp paid off, and he knocked the collective socks off his audience, including Tony: "I recollect that he was very British and proper, and amazingly poised and confident for a graduate student—he held his own at the UCLA meeting, and at a meeting in Kyoto he came to instead of Mike in November 1985." Julian doesn't recall being particularly worried by the attention: "It was just exciting, really. I didn't feel out of my depth—I feel much more out of my depth now than I did then. At the time I felt I knew more about this than anyone else in the world and I think that was probably true."

In May 1984, the Waterfield lab had another *Nature* paper to celebrate, this time with a crowd of collaborators from the Ullrich and Schlessinger labs, in which the complete coding sequence of the human EGF receptor gene was presented. Comparison of the sequence of the cellular gene with its oncogenic relative v-*erb-B* showed that the oncogene was truncated, deleted, and mutated in such a way that the protein it encoded was no longer regulated by EGF binding, but instead was permanently switched on, pushing cells to grow uncontrollably. This, together with the PDGF story, was the clinching evidence for a truth that is now ingrained in the collective consciousness of cancer biologists: oncogenes cause tumours because they encode mutationally activated components of normal cellular growth pathways. As befits their status, the three papers that the Waterfield lab published in that one remarkable year have been cumulatively cited around 6000 times. The perfect combination of state-of-the-art database mining, protein purification, and sequencing, matched with Mike's outstanding eye for a good problem and a good student, had paid off in spades.

In cancer, there are now many examples of mutated or overactive growth factors and receptors causing particular human tumours, and numerous drugs have been developed that target such molecules, many based on neutralising monoclonal antibodies such as Peter Goodfellow's EGFR1. Mutations in the EGF receptor itself have been found in multiple cancer types, and several effective EGF receptor–inhibiting drugs are now in use in the clinic. One of the best known of these is the therapeutic antibody Herceptin (trastuzumab), a standard and effective part of first-line breast cancer treatment developed at Genentech by, amongst others, Axel Ullrich.

Thanks in part to the Waterfield lab and their collaborators, signal transduction (how extracellular messages affect intracellular events) became the fastest growing field in biology in the 1980s. However, although it was evident that deregulation of growth factors and their receptors could cause cancer, the pathways leading from receptor activation into the cell petered out in a dense thicket of uncertainty. One obvious line of inquiry, trying to work out what proteins the receptors were able to phosphorylate, proved to be annoyingly circular, because their only convincing substrates were themselves. Educated guesswork, seeing whether other known oncogenes could be activated by growth factors, was deployed with more success. In 1984, a paper by Tohru Kamata and Jim Feramisco gave the first vague hints that addition of EGF to cells somehow stimulated the activity of the Ras oncoprotein. This tenuous link strengthened into reality in 1986, when Dennis Stacey and colleagues used antibodies blocking Ras activity to prevent transformation by receptor tyrosine kinases. Ras clearly had something to do with receptor signaling, but whether it lay directly downstream from the receptors or was on a parallel pathway was still an open question.

Ras had a great deal of previous form as an oncoprotein. In 1979, Bob Weinberg's lab at the Whitehead Institute in Boston showed for the first time that there were human genes that could cause cancer when mutated. Using the new technique of DNA transfection, whereby foreign DNA could be introduced directly into cells, they showed that a fragment of DNA from a human tumour was able to transform mouse NIH3T3 tissue culture cells. Other groups began to see the same results using DNA from other tumour types, and together, they realised that the fragments of human DNA they were isolating all contained the H-*ras* gene, the cellular homologue of v-*H-ras*, one of two closely related retroviral oncogenes from rat sarcoma viruses. Ras family proteins are attached to the inside of the cell membrane and belong to a larger protein family, the GTP-binding proteins. Such proteins are able to bind guanine triphosphate (GTP), a product of cellular metabolism, and convert it to guanine diphosphate (GDP). It was known that the mutations that converted normal Ras proteins into oncogenes caused them to stick in the GTP-bound activated state so that they were switched on the whole time, but here the trail ran out. Ras sat in solitary splendour, with its upstream and downstream partners still shrouded in mystery.

In 1987, after writing up probably the most spectacular PhD ever done at the ICRF, Julian Downward was sucked into the black hole surrounding the Ras proteins when he started a postdoc in Bob Weinberg's lab. It was fortunate that he'd had such a good experience as a PhD student, because his postdoctoral work was nowhere near as successful, consisting of mostly fruitless attempts to work out what the trigger for activation of normal Ras proteins might be: "I spent most of 1987-9 throwing EGF and similar growth factors at cells and looking at the GTP bound to Ras, but never saw any effects, and many others had the same experience. It was a technical problem and in fact was not completely solved for a long time … the stimulatory effects were just too small in most of the cell systems to rise above the noise inherent in the assay."

In 1989, Julian was recruited back to the ICRF to run a laboratory in his own right, and arrived still shackled to his annoyingly intractable postdoc project. Although Mike had moved on to become Director of the Middlesex Hospital branch of the Ludwig Institute for Cancer Research, Julian still found a sympathetic audience for his Ras-related problems: his fellow group leader Doreen Cantrell, who was (and still is) a leading T-cell immunologist. T cells are a vital part of the immune system, binding to foreign antigenic invaders and activating the body's powerful defences against disease. As an experimental tool for studying signal transduction, they are also extremely attractive because they can be easily purified and fooled in vitro into activating themselves if they are fed mock antigens. Doreen suggested to Julian that her T-cell system would be an ideal place to look at what was activating Ras. The shift in perspective, as can often happen in scientific research, broke the logjam. In just four months, the collaboration succeeded in doing what Julian had been labouring over in Boston. Working together, the two labs showed for the first time what everyone had suspected but had been unable to prove: Ras could be activated rapidly and strongly in response to extracellular signals stimulating a receptor, in this case, the T-cell receptor, the cell surface molecule on T cells to which antigen binds.

Once Ras activation in T cells had been shown, it could be used to tune up the problematic GTP-binding assay to detect what else switched

Ras on. The signal-to-noise ratio was improved to the point that many other receptors were shown to turn Ras on, including Julian's old friend the EGF receptor. The race was now on to join up the dots between the receptors and Ras. Julian and his postdoc Laszlo Buday realised that work performed genetically dissecting Ras signaling in fruit flies could be put together with biochemical clues from mammalian cells to produce two likely candidates for the missing links, the Sos and Grb2 proteins. Sos (Son Of Sevenless; the *Drosophila* field is inordinately fond of giving genes daft names) was known to act downstream from the fly EGF receptor gene and upstream of fly Ras, and there was also good reason to think that it was a guanine exchange factor, able to catalyse rapid exchange of GDP for GTP, and therefore a prime candidate for activating Ras. Grb2 had been proposed to be an adaptor protein, acting as a link between the EGF receptor and intracellular signalling components, and there was indirect evidence implicating it in Ras activation. Leaving T cells for fibroblasts, Julian and Laszlo wallowed enthusiastically in radioactivity for a few months, labelling cells and using antibodies to try to immunoprecipitate Ras protein along with Grb2 and Sos, and then doing in vitro biochemistry to show that the interactions they detected resulted in Ras becoming activated. Their efforts successfully showed that the EGF receptor, when activated by EGF binding, recruited Grb2 to its intracellular tail, that Grb2, in turn, bound Sos, that Sos had hold of Ras, and that all together, this enabled Ras to be activated. This work, together with a similar paper from ICRF alumnus Tony Pawson's lab in Toronto, defined the missing upstream links.

But what lay downstream from Ras, between it and the genes controlling the cell cycle? Part of the answer lay with another proto-oncogene, c-*Raf-1*. Raf1 protein is a kinase, and there was indirect evidence linking it with activation of Ras, because disruption of Raf1 activity had been shown to block Ras action. The Downward lab used a piece of Raf1 as a biochemical bait to see whether it was able pull down Ras protein and showed that, indeed, Ras and Raf1 interacted, but only when Ras was in its activated, GTP-bound state. This was a crucial interaction to map, because downstream from Raf1 lay the mitogen-activated protein kinase (MAPK) cascade. Kinases belonging to the MAPK family are used throughout evolution to control the cellular responses to external signals such as growth factors, nutrient status, and stress or inductive signals, and MAPK had just been shown by Richard Treisman's lab, just upstairs from Julian at the ICRF, to regulate transcription factors. The extracellular signalling pathway had reached the nucleus, and ICRF scientists had played a big part in illuminating many of its twists and turns.

There is one further chapter to the story, as became clear to Julian that not all of the things that Ras was known to be able to do could be explained by it simply activating the MAPK pathway. Another protein, phosphatidylinositol-3-kinase (PI3K), had been shown in 1992 to interact with Ras in vitro, and PI3K was known to be activated in response to the same stimuli as Ras. Julian's lab, in collaboration with Mike Waterfield's at the Ludwig Institute, showed in 1994 that activated, GTP-bound Ras bound directly to PI3K, and that this interaction was likely to result in PI3K being activated in turn. The work placed Ras at the apex of two downstream signalling cascades, both of which are vitally important for cell growth.

The basic mechanism of how extracellular signalling works has been fleshed out in the intervening years such that we now have a very detailed map of signal transduction in both normal and cancerous cells. Growth factors and other extracellular stimuli sit at the top of a host of pathways governing cell growth and differentiation, and the molecular wiring diagrams for different cell types are varied and complex. Scientists are turning towards systems biology to model how perturbations in one part of a cell's wiring change the outcome for the cell, and signal transduction research is becoming rigorously quantitative. Fresh insights into the processes involved are won at huge cost; there is no low-hanging fruit left to pick in the garden of molecular biology. The excitement of the early days, when huge leaps in understanding were almost commonplace, has gone, to be replaced by the tramping feet of an army of scientific

Mike Waterfield and Julian Downward, 2002.
(Photograph courtesy of CRUK London Research Institute Archives.)

foot soldiers marching towards the ultimate goal—complete knowledge of a cell and the medical exploitation of that knowledge. They will get there, and there will doubtless be other kinds of fun along the way, but there will never be another time when one lab equipped with a small computer, a fancy sequencing machine, and some great ideas could change our view of the molecular landscape so profoundly.

Further Reading

Cohen S. 2008. Origins of growth factors: NGF and EGF. *J Biol Chem* **283:** 33793–33797.
A very clear account of Stan Cohen's early life and work.

Levi-Montalcini R. 1988. *In praise of imperfection*. Basic Books, New York.
Rita Levi-Montalcini's autobiography. A remarkable book about a remarkable woman

Web Resources

http://blip.tv/portland-press-ltd/mike-waterfield-ludwig-institute-for-cancer-research-ucl-london-2691775 Mike Waterfield in action.

www.asbmb.org.au/magazine/2002-August_Issue33-2/Pehr%20Edman%20Feature.pdf Contains an *extremely* frank account of Pehr Edman's personality by his ex-research assistant, as well as an assessment of his life and work.

www.youtube.com/watch?v=wiu7DMjG6RE Russ Doolittle is an enchanting speaker. Watch him lecture.

Quotation Sources

BBC TV *Horizon* series. *Cancer, the pattern in the genes.* Broadcast 5 December 1983.

Ibid.

Ibid.

Levi-Montalcini R. 1988. *In praise of imperfection: My life and work.* Basic Books, New York.

Newmark P. 1983. Oncogene discovery. Priority by press release. *Nature* **304:** 108.

William Dreyer interview by Shirley K. Cohen, 18 February–2 March 1999. Oral History Project, California Institute of Technology Archives, reprinted with permission. http://www.archives.caltech.edu/search_catalog.cfm?results_file=Detail_View&recsPerPage=1&firstRecToShow=72&search_field=&entry_type=NonPhoto&photo_id=&cat_series=oral%20history.

CHAPTER 6

Divide and Rule

When talking about his earlier life and times, Sir Paul Nurse, President of the Royal Society and Nobel laureate, uses one word rather frequently. It is, somewhat surprisingly, *failure*. A clever grammar school boy from a working class family, it took him a while to get to grips with passing exams at school, and his route into university was more tortuous than most. He did an unexciting PhD, publishing two papers, which in the years since have garnered only five citations. After his postdoc, working on an untrendy subject, he failed to get any of the lectureships for which he applied and had to take a post funded by short-term money at the University of Sussex, to keep afloat. When that money ran out, he failed to get still more jobs and was on the brink of leaving the country for a laboratory in Germany when he was rescued by Walter Bodmer, Director of the ICRF.

In addition to this lexicon of failure, other facts fall out from past interviews with Paul. How did he get interested in science? In two ways: first, as a small, solitary boy, standing entranced in the street in his pyjamas as Laika the space dog and Sputnik 2 flew through the night sky over Wembley; subsequently, observing with fascination the life beneath his feet as he walked to and from school through the fields. Interviewers have covered other subjects as diverse as his Desert Island Discs (luxury: a telescope; favourite record: Billy Joel "Just the way you are"), his attempts to learn the recorder as an adult, and why he shaved off his moustache ("I thought it would make me look more vulnerable."). In recent years, the bombshell of Paul's discovery that his parents were, in fact, his grandparents, and his real mother was the person he had thought was his older sister, has added an extra level of tragic complexity to the story.

These snippets are, of course, nowhere near the whole truth, but they do supply some intriguing clues to what makes this apparently bumbling (*mess* is another word that recurs frequently) science enthusiast tick and how he became what he is today. And one of these clues is the constancy of the narrative itself. Science is too frequently disseminated to the public as a series of facts ("scientists have discovered that …"), and it is only rarely that the story behind the facts is regarded as relevant. But for Paul's research, the narrative is everything, and what he did to win his Nobel Prize, what he is still doing today, is living out a scientific story that he has been telling since he was a young man in his 20s.

Paul Nurse's baptism as a scientist came in a most unlikely place, just off the North Circular Road near the Hanger Lane gyratory in Willesden, North London, at the Guinness Brewery-owned Twyford Laboratories. He was 17 years old and had finished school a year early, leaving with good A levels, but crucially lacking the French O level he would need to get him into university in the 1960s. His school career had been almost comically inconsistent; he veered between the bottom and the top of the class, sometimes in the same subject, but his bête noire was definitely languages. Despite having a razor-sharp mind able to grasp almost anything as long as it has logical consistency, Paul struggled, and still struggles, with pronouncing unfamiliar words in English, let alone in French. Combined with a total ignorance of grammatical structure and an inability to associate the strange French words with either their English equivalents or the objects they described, he was a walking linguistic disaster. Accordingly, the original plan for his post–A level year was to stay on in the sixth form to work at his French and to redo an A level or two (an A and two B grades not being considered quite good enough). However, after a term of boredom and wanting to earn a bit of money to help with the family income, Paul dropped out. Resigning himself to taking French O level by himself (famously, he eventually failed it a grand total of six times), he got a job making biological media for a group studying *Salmonella* food poisoning at the Twyford Labs, while he reapplied to university.

Paul's boss at the Twyford Labs was Vic Knivett, a shy but very kindly man with an incongruous interest in Russian dancing, of which he was an enthusiastic exponent. The new junior technician Nurse once caught him practising after hours on one of the benches in the lab, much to Knivett's embarrassment. The job was undemanding, consisting of visiting the inhabitants of the lab every morning to take their orders for that day's media, mixing it up from powders, and then delivering it. Paul realised quite quickly that because everyone always asked for the same items, he could make solutions up in bulk on Mondays, store them in the cold room, and then distribute them upon request. This left the problem of what to do for the rest of the week, so he approached a rather surprised Knivett to ask him for more work. Knivett, to his credit, recognised a gift horse when he saw one and put Paul onto a proper research project, looking at rapid ways of detecting *Salmonella* by using the new technique of fluorescence microscopy. Paul thrived on the work, and for the next few months was happily occupied during the day, although he still had to wrestle with intractable French vocabulary in his spare time.

Fortunately for the young Nurse, somebody high enough up in the hierarchy of a British university had the good sense to realise that incompetence in a foreign language was no barrier to becoming a good scientist. In 1967, after a journey on his motorbike up the new M1 motorway for a hastily convened interview, Paul started a degree in Biology at Birmingham University, thanks to John Jinks, head of the Genetics Department. Jinks's own staunchly working-class background may have predisposed him towards bucking the Establishment's desire for unnecessary qualifications. He cooked up a deal with a member of the French faculty to allow Paul to take and pass a semi-fictitious course in French in his first year, and, with the rules bent but not quite broken, Paul was free to start on his path to scientific glory.

The path certainly didn't seem particularly glorious to begin with. Paul is remembered by Jack Cohen, one of his Birmingham lecturers, as being undistinguished, and Paul himself summarises his third-year undergraduate project, on the respiration rate of dividing fish eggs, as "a disaster, but a useful disaster," falling at the last hurdle because a vital control had been omitted. Having started off with an interest in ecology and genetics, he abandoned them early on, the former because he "couldn't bear to be muddy and wet and out of control" and the latter because of an excessive emphasis on cytogenetics, not the most exciting of topics. Instead, Paul took a classical biology degree, majoring in botany and zoology, and got a First, despite having missed all of the second-year zoology lectures in a valiant attempt, no doubt, to fail at something yet again.

The haphazard route by which he arrived at his first-class degree had provided Paul with two unusual mental resources, which give some clues to his subsequent outstanding success: he was very relaxed about failing, never regarding it as much of a problem, and his sense of intellectual excellence was entirely self-referential, making him highly competitive, but only with himself.

Here is Paul on the benefits of early failure:

> Because I failed at so many things so often, because I was in a mess at school ... it gave me a sort of internal discipline—you take less note of what other people think of you, what other people say, because you don't get off on being praised about things. I had to be resilient inside. ... I was constantly comparing me to me when I did well, and not with other people. I realise it's very odd, but it's really useful, because when I ... failed examinations, I couldn't get into university, I couldn't get a job, when you put all that together, it was a constant low to medium level [of] failure about things. So when I got to difficult problems and I failed, I didn't go into depression or anything. And when you get into research, it's constant failure all the time, and I was perfectly trained for it.

So what does a self-competitive optimist with no fear of failure do for a PhD? Probably influenced at least slightly by the fish eggs debacle in his final year, Paul had already decided that there was no point slogging away in a lab unless you were working on something that really mattered, something that was important enough to stand a good risk of failure on a grand scale. Paul's lecturers at Birmingham, traditional biologists all, had encouraged him to apply to botany departments for his PhD, and, although he had been offered a project at Cambridge, the topic, on the biochemistry of RNA, was not to his taste. His omnivorous reading habit had led Paul far beyond the mundane nuts-and-bolts problems that such a PhD project would address, into a theoretical land of conjecture and speculation about how biological entities are ordered; how does an organism build itself up from a single cell, and how does the resulting community of cells interact during a lifetime?

Forty years on, Paul is still preoccupied with this question, which lies at the heart of understanding biology itself, and the length of elapsed time is indicative of the scale of the problem. Wisely, his younger self made a very logical decision to start small, studying one of the most important but probably the simplest developmental decisions a cell can make—how its structures are organised in space and time to enable it to divide into two. Understanding the molecular events that regulated this decision, dissecting the changes in enzyme and gene regulation that were required, might just be approachable at a single-cell level, even taking into account the rather limited molecular techniques then available.

Cell division occurs when a cell has completed one round of the cell cycle, the period between the formation of a cell by division of its mother cell and the time when the cell itself divides to form two daughters. Paul quickly realised that regulation of the cell cycle was pretty much a black box, ignored by the developmental biologists of the time. Cell cycle research was the preserve of a small band of zoologists, and what was known was essentially descriptive: patient observations made over many years by staring at cells under microscopes and carefully drawing what could be seen. Because the most spectacular part of the cell cycle is mitosis, the point at which one cell splits into two, there were plenty of drawings of that, and researchers were very interested in the mechanics of how the split happened, but the rest of the cycle, where the cell just sits there looking boring, was mostly unexplored. In the late 1960s, a few biochemists had realised that the ostensibly boring bit of the cycle was, in fact, the time when the cell had to work hard to double all of its contents, preparing for the mitosis that would split these contents between two daughter cells. Some work had started to look at what was going on there, but the field was very much in its sleepy infancy with a small number of people doing mundane things.

After his uninspiring visit to Cambridge, Paul eventually settled on a PhD at the University of East Anglia (UEA) in Norwich. The Biological Sciences School at UEA had a lecturer, Tony Sims, who was just beginning a project on cell cycle changes in the enzymes involved in amino acid metabolism in the yeast *Candida utilis*, and this seemed to fit Paul's criteria for fundamental cell cycle research very well. In addition to the suitability of the project, Norwich as a place was attractive to Paul, partly because his family originated from there and it was where he had been born, and partly because as a budding socialist, he was very taken by the idea of working at a new university, free from the Establishment constraints that he'd already had to fight against to even get a degree. UEA was less than 10 years old and had taken part enthusiastically in the period of student unrest sweeping through Europe in the late 1960s; the unfortunate Princess Margaret had had a blazing Union Jack thrown at her during a visit a couple of years before Paul's arrival. Paul signed on and started work there in 1970.

The Norwich PhD was a hard slog, with very little to show for it—a "baptism of learning by failure" in Paul's words. Tony Sims was a very good supervisor, who gave Paul a thorough grounding in the craftsmanship of experiments, but the experiments themselves were frankly tedious. Far from shining a light on how the cell cycle was regulated, Paul found himself growing up vats of yeast cells, concentrating them by centrifugation to make smelly beige pellets, and then measuring the amino acids and proteins he could extract from the pellets. He thought about other careers, perhaps studying the philosophy or sociology of science, but in the end, the camaraderie of the Sims lab, where drudgery was enlivened by the sympathy and support of his fellow investigators, carried him through. By the end of the second year in Norwich, Paul had decided to stay as a practising scientist but was starting to fret about what to do next for his postdoctoral work. He was still completely committed to working on regulation of the cell cycle, but he knew now that biochemical approaches were not likely to contribute many answers to the problem he wanted to solve.

His PhD work on amino acid metabolism, coupled to his bookworm tendencies, had led Paul into thinking a lot about metabolic pathways, and particularly the ways of controlling flux through a pathway. Metabolic pathways are the production lines of the cell, combining raw materials into more sophisticated molecules, which are then fed into the vast machinery required to maintain and drive the life of the cell. However, as for any production line, the rate of flow through the system must be controlled very carefully, in order to ensure that products are made in the right amounts, neither flooding nor starving the market, so to speak. Metabolic pathways can be regulated either at multiple points or in just a few places, depending on how they are set up, and it occurred to Paul that this sort of reasoning could be applied to cell cycle control. If the cell cycle were regulated in only a few places, then it should be possible to speed up the cycle drastically, or alternatively slow or stop it altogether, by messing with just one regulation point. To switch similes, if the cell cycle controls were like the sluice gates in a watermill, the speed at which the cell traversed the cycle, like the speed at which a mill wheel turns, could be entirely regulated by tinkering with just one item, the molecular equivalent of opening and closing a sluice. If this were true, then there should be cells that went into mitosis too early or went into mitosis late or not at all, and by finding the mutations that had caused the changes, the identity of the regulators could be established.

This theoretical daydreaming was all very well, but there was one big snag to it; much as he racked his brains, Paul was unable to think of how to approach the problem practically, how to find the mystery regulators of the cell cycle. Then, one night in 1972, staying late to do yet another boring experiment, he found the answer. It lay in a paper in the journal *Proceedings of the National Academy of Sciences*, written by an American called Leland Hartwell.

If Paul could be said to have worked his way down to yeast from a starting point of larger organisms, Lee Hartwell had done the opposite, coming up from the world of bacteriophages and viruses, where he had been involved in some of the earliest studies on how genes were regulated. Having decided somewhat prematurely that gene regulation was pretty well understood by the mid-1960s, Hartwell had set himself a new research problem—how cell growth was regulated. In looking for a suitable eukaryotic model in which he could apply the principles of bacterial genetics that he had learnt as a student, he was introduced to budding yeast, *Saccharomyces cerevisiae*, which at that time was one of the only single-celled eukaryotic organisms with easy genetics. It was possible to treat budding yeast cultures with an unpleasant chemical called nitrosoguanidine and then test the surviving yeast cells to see if their DNA had mutated, using a straightforward and quick assay, temperature sensitivity. The principle was extremely simple; mutated DNA would in some cases lead to mutated protein, and some mutant proteins cannot function properly at the wrong temperature. Such mutations can therefore be picked up by growing duplicate plates of cells, one at the right temperature and the other at the wrong, or restrictive, temperature, and then comparing the plates to detect changes or deaths at the restrictive temperature.

Using this method, Hartwell quickly isolated about a thousand mutants that were unable to grow at 36°C but were quite happy at normal room temperature, and began to categorise them based on their defects. Although some were too boring to study, he did find mutations that could be specifically attributed to defects in one of a number of cellular processes such as DNA, RNA, or protein manufacture, cell division, and cell growth. To prove the point that this genetic approach would work to identify the individual genes responsible for each mutation, Hartwell used his protein synthesis mutants to show that three were caused by specific defects in three particular proteins, and hence the genes that encoded them.

Hartwell's entry into the cell cycle field was precipitated by one of his PhD students, Brian Reid, who noticed something very odd about some of the mutants. Budding yeast, as the name suggests, divides by growing a bud from the mother cell, with the bud eventually getting big enough to pinch off from its parent to form a new daughter. In a normal culture, all of the cells are at different points in the cell cycle, and, therefore, they all look different. However, looking down a microscope one day, Reid saw mutants that, when shifted to the restrictive temperature, all started to look the same—their buds were exactly the same size. He and Hartwell realised that they must all be stuck at the same position in the cell cycle and that, therefore, mutations affecting the cycle directly could be identified very easily, simply by looking at the cultures. Furthermore, because the size of the bud was an indicator of where exactly in the cycle the yeast was, mutants could be subclassified depending on bud size into those affecting different steps of the cycle.

Using time-lapse video microscopy, Reid, Hartwell, and a second PhD student, Joe Culotti, then went one step further. By filming mutant cultures continuously, starting at the time at which the cells were shifted to the restrictive temperature, they found that some cells stopped in the cycle almost immediately, whereas others went around another cycle before arresting at the same point. This meant that mutations could act at one particular point in the cycle and that if cells happened to have got past that point before the temperature shift, they would continue until they ran up against the mutation point in the next cycle. Careful tracking of the cells able to do another round of cycling also threw up another interesting observation—mutations did not necessarily act at the point where the cells arrested in cycle but could exert their effects much earlier on.

It was Hartwell, Culotti, and Reid's paper describing this work that Paul fell upon during his nocturnal trip to the UEA library, and to someone searching for a way to get a grip on an intractable problem, it was an astonishing moment: As Paul says, "It completely blew my head off." The more Paul discovered about Hartwell's work, the more he became convinced that it was the simplest, most direct way to get to the heart of what regulated the cell cycle. The mutants would lead straight to the genes of most importance, and then identifying the genes and working out what they did would put the framework in place to solve the whole problem. Genetics, not biochemistry, was clearly the way to go. Fired with evangelical zeal, Paul began working out the steps that would enable him to follow Hartwell into the maze.

To do a postdoc studying cell cycle regulation in yeast, Paul was going to need several things. He knew something about yeast already, so that was a start, but he knew nothing about genetics, and he also had to find a postdoctoral supervisor who would allow him to work unimpeded on the project, without trying to foist their own interests on him. Paul briefly considered applying to Hartwell's lab, but he and his wife Anne had no wish to go to the United States, and Anne, who had recently finished teacher training, was keen to stay in Britain and start work. This threw up a huge problem; the main person working on the cell cycle in yeast in the United Kingdom at that time was Murdoch Mitchison in Edinburgh, but he had no expertise in genetics and additionally worked, not on budding yeast, but on a very distantly related cousin, *Schizosaccharomyces pombe*, fission yeast. Undaunted, Paul wrote to him and went up on a bright, snowy winter's day to Edinburgh for an interview.

Murdoch Mitchison, part of the "third of a ton of Biology Professors" born to the writer, poet, and activist Naomi Mitchison, and older brother of Av, David Lane's mentor, was very definitely interested in cell division, having recently published an extremely influential book, *The Biology of the Cell Cycle*. Mitchison had shifted into the cell cycle field in the 1950s, after earlier work on sea urchins and red blood cell membranes, and had realised early on that fission yeast was a great model, perhaps better than budding yeast, because to divide, it simply got longer and longer until it split in two. This meant that position in the cell cycle could be measured by length, rather than the more subjective bud size. The fission yeast cell cycle also looked a lot more "mammalian" than that of budding yeast, because the timing of each part of its cycle was similar to that of higher eukaryotes, a fact that fueled Paul's growing enthusiasm for his potential new laboratory workhorse.

Mitchison had thought a great deal in biochemical terms about how the cell cycle might be regulated, and *The Biology of the Cell Cycle*, although a dry and dusty read, was the first clearly to lay out two new and important concepts. First was the relationship between the cell cycle and cell growth, the idea that because cells can only divide if they are big enough, they must therefore have a way of sensing size and tying that into cell cycle control. Second was the recognition of the causal dependency of events in the cell cycle, that one event had to happen in order that another, subsequent one could occur. Furthermore, Mitchison proposed that the time it took for a cell to complete the different steps in the cycle could be determined by these dependencies, and also by external, master timing mechanisms. This idea was a forerunner of more advanced theories of checkpoint control, which underpin our modern understanding of the cell cycle.

Paul, accustomed as he was to thinking about cell cycle control in splendid intellectual isolation, recognised a kindred spirit, who clearly had no regard for other people's opinions, but simply wanted to write a book about something in which he was really interested. His trip to Edinburgh confirmed his suspicions. Mitchison happily spent an entire day with him talking about what Paul might do as a postdoc and, furthermore, acknowledging his own ignorance of genetics, suggested that in order to get to where he needed to be, Paul should spend a few months in Bern learning some tricks of the trade with Urs Leupold, the father of fission yeast genetics. Finally, Paul had found a decent system, a decent problem, a logical way of thinking about it, and a great place to work. He and Anne left Norwich in 1973, bound for Bern and then Edinburgh.

In Bern, Urs Leupold, a brilliant but eccentric geneticist, spent hours with Paul to teach him the rudiments of genetic analysis, although Paul remembers some of his wilder ideas as quite "barkingly mad." Paul took to the abstract, highly theoretical subject material like a duck to water; his great strength had always been that he was almost pathologically orderly in his thought processes, and the rigorous logic of theoretical genetics matched his mind in a very satisfactory way. As well as the long hours with Leupold, Paul began to work with the lab's fission yeast strains and set up his first temperature-sensitive screens, looking on the replica plates grown at the restrictive temperature for cells that were getting too long because they couldn't complete the cell cycle and split into two. By the time he moved back to Edinburgh, and in the months following, he had isolated about 30 different cell division cycle (*cdc*) mutants. The Mitchison lab's expertise in physiological studies of the cell cycle meant that he could categorise the *cdc* mutants on the basis of where they were blocked, just as Hartwell's lab had done with budding yeast, and Paul found mutants blocked for DNA synthesis, mitosis, and cell division, aided in subsequent work by Kim Nasmyth, then a new and "frighteningly bright" graduate student in the Mitchison lab.

Despite his being the genetic cuckoo in a very biochemical nest, Edinburgh as a workplace was all that Paul had hoped for. Mitchison gave Paul complete freedom to do what he liked and was interested enough in Paul's new approach to the cell cycle to talk to him endlessly. The lab was small, but Paul's colleagues were smart and committed, and above all, the science there was always led by Mitchison's belief that passionate interest in a subject is far more important than whether it is fashionable or not. In this very congenial environment, Paul began to lay the foundations for all of his subsequent work on the cell cycle. His first big intellectual breakthrough, however, was not the product of the theorizing of a finely honed mind, but a simple piece of serendipity brought on by a desire to cut some experimental corners.

The temperature-sensitive screens that Paul had been performing were proving to be very fruitful, but searching through thousands of colonies on agar plates looking for the small number that might have mutated was unexciting and long-winded. In a bid to speed matters up, Paul decided that a better approach might be to mutate the cells, grow them up at the restrictive temperature, and then separate them on the basis of their size; the mutant cells stuck in cycle would be much larger, and they should be easy to separate by spinning through a lactose gradient in a centrifuge because the largest cells would all collect at the bottom of the gradient. The big cells could then be recovered by growing on agar at a normal temperature, and then tested by temperature shift, as normal. In fact, the idea was a complete failure for what Paul wanted, because most of the mutant cells didn't recover, and the normal cells swelled up too as a result of the stresses they were under. However, whilst fruitlessly looking for extra-long cells extracted from the bottom slice of the gradient, Paul noticed to his surprise that there were some strange-looking, small, round cells that had made it to the bottom slice because they had clumped into a ball, which to a centrifuge looks just the same as a big cell. He was about to throw the plate containing the tiny cells away but suddenly stopped. If cells that were too long were caused by a block in the cell cycle, then surely, cells that were too small might be caused by a cell cycle that ran too fast, dividing before the cells had grown enough? Long cells, he realised, were not the key to finding the rate-limiting control steps in the cell cycle, because any event necessary for the cell cycle would become rate-limiting if it were inhibited. However, if there were a mutant that speeded up an event rather than inhibited it, such a mutant would be in a bona fide rate-limiting gene, because speeding up a normal rate-limiting step would push the cells to cycle faster. It was the difference between getting a puncture in a bike tyre (necessary), which would cause a journey to be slowed, and pedaling faster (rate-limiting) to arrive earlier.

Although Paul was right, and the *wee* mutants (christened in acknowledgement of their Scottish origins) played a vital part in finding the key regulatory molecules of the cell cycle, his sudden flash of intuition did not get the reception he'd hoped for: "I was so excited by this, so excited! I went and showed people, but nobody quite got it. It was so frustrating, because I couldn't understand why they just didn't see that it was important. In fact, some people thought it was budding yeast, that I'd contaminated the culture, which I knew I hadn't. It did make me realise that you do have to be very open-minded, see what you get and capture it."

Paul's 1975 *Nature* paper on the *wee* mutant phenotype, "Genetic Control of Cell Size at Cell Division in Yeast," saw his arrival on the yeast cell cycle scene, and during the next few years in Edinburgh, he published another 15 papers characterising in detail the stable of cell cycle mutants that he had built up. At first, his work ploughed a very similar furrow to Hartwell's, and Paul worried that he would always be running to catch up the budding yeast cell cycle community, which was much larger than the small band of fission yeast enthusiasts. However, Hartwell's work ran into a slow patch, and without his input, budding yeast cell cycle work stalled for a few years, allowing the fission yeast community to draw level.

The cell cycle is split into four phases: G_1, S, G_2, and M. S phase is the time during which DNA is duplicated; M phase is mitosis, when the duplicated DNA is split and the cell divides into two; and G_1 and G_2 are the gaps on either side of S phase. The transitions between G_1 and S phase and G_2 and M phase are key regulatory points in the cell cycle, where go/no go decisions are made regarding DNA synthesis and cell division. Paul's work in Edinburgh, done in collaboration with Peter Fantes, Kim Nasmyth, Pierre Thuriaux and others in the Mitchison lab, focused mainly on the G_2/M transition, which is the preeminent control point in fission yeast. By the late 1970s, they had shown that all of the *wee* mutations Paul had isolated were causing changes in just two genes, called *wee1* and *cdc2*, and that these two genes were rate-limiting for the G_2/M transition, somehow telling the cell cycle that the cell was big enough to divide into two without stinting on the genetic and biochemical dowry of either daughter cell. Interestingly, the *cdc2* gene had first been identified as a mutant in the elongated cell screen, and Paul and Pierre Thuriaux proposed in 1980 that *cdc2* could be mutated either so that it lost its function, in which case, mitosis was blocked and cells elongated, or so that it became more active, in which case, cells divided too early and were too small. They further suggested that *cdc2* and *wee1* were working together in a network to regulate the G_2/M transition and hence the onset of mitosis.

Although G_2/M was the point of interest in the cell cycle for fission yeast biologists, this was not the case for their budding yeast colleagues. Budding yeast barely has a G_2 phase, moving almost immediately from S phase into M phase, and the major control point is at the G_1/S transition, christened "Start" by Hartwell. Start also had a counterpart in animal cells; in 1974, Art Pardee had proposed what he called the "restriction point" in the mammalian cell cycle. This was the point at which animal cells made a decision to divide or not based on the supply of nutrients, and it occurred at the same time as Start: at the G_1/S transition.

The buzz going round about G_1/S and Start meant that there was not a lot of interest in G_2/M control, and Paul, by now looking for a lectureship somewhere as a prelude to leaving Edinburgh, made a pragmatic decision to do a little bit of opportunistic work on the G_1/S transition in fission yeast, in order to look more attractive to prospective employers. He and Yvonne Bissett, a technician in the Mitchison lab, set up a quick experiment to see whether any of their temperature-sensitive *cdc* mutants had anything to do with Start. The assay was simple, based on a normal lifestyle decision made by yeast cells, which divide happily in conditions of optimal nutrition, but stop dividing and mate (conjugate) when they detect problems with food supplies. The point at which cells make the decision to divide or to conjugate is Start, and once a cell has passed Start and is committed to divide, it cannot conjugate. Therefore, at the restrictive temperature, if a *cdc* mutant were needed for Start, cells would arrest in G_1, would not be able to commit to entering S phase, but would still be able to conjugate. If the *cdc* mutant were irrelevant to Start, cells would arrest at some other point in the cycle and no conjugation would occur. The readout of the assay was to select for cells that could conjugate, which Paul and Yvonne did by rigging the medium on the agar plates they used such that only conjugated cells would stay alive.

The experiment was set up, and Paul chose as a negative control a *cdc2* temperature-sensitive mutant—because *cdc2* was a G_2/M regulator, cells should arrest in G_2 and would never conjugate. However, a gremlin seemed to have sneaked into the Mitchison lab; to Paul's intense frustration, the seemingly simple control just would not work. Every time he and Yvonne ran the experiment, the *cdc2* control plates always had a few surviving, conjugated colonies and were therefore very unsatisfactory controls, meaning that the rest of the data from the experiment could not be trusted. Paul tried two different *cdc2* mutants and repeated the experiment time and time again, but always got the same result. What on earth was going on? As Paul says: "I thought and thought and thought. Of course I assumed I was stuffing it up. All the usual things—I thought the temperature wasn't right—you go back to the waterbath with the thermometer. And then it came to me that it had a double block point."

The "double block point" got the paper into *Nature*. What Paul had realised was that *cdc2* might be not only the rate-limiting regulator for G_2/M, but also the key to Start control at G_1/S. Viewed in this light, the data made perfect sense; when shifted to the restrictive temperature, *cdc2* mutant cells could either arrest in G_2, in which case they'd be unable to undergo conjugation because they were in the wrong bit of the cell cycle; or just before Start, in G_1, where they could conjugate with ease. This would appear as a dribble of surviving colonies at the end of the assay, just as Paul and Yvonne had seen.

Paul again: "That paper is one of my favourites because to get to that conclusion was not intuitive—it wasn't how people were thinking about it; it wasn't how I was thinking about it. To make the interpretation, to say, 'Well, I can test *that* if I do *this*, but the chances of this working are zero.' To do that, and then to get the result.... I don't talk about it so much these days but it's one of the three or four things I'm most pleased with."

The *Nature* paper, "Gene Required in G_1 for Commitment to Cell Cycle and in G_2 for Control of Mitosis in Fission Yeast," was published in August 1981, by which time Paul had moved on from Edinburgh, to a three-year position at the University of Sussex in Brighton. Despite his attempts to up his trendiness factor, he had been unable to get a lectureship anywhere and had had to settle for a short-term fellowship. Although not very secure in terms of job prospects, the location of Paul's new lab was in one way ideal; the School of Biological Sciences at Sussex had expertise in what he was planning next—molecular biology.

Up until now, all of Paul's work with fission yeast had involved the isolation and characterization of mutants that were assigned gene names but remained theoretical entities. The genes existed only on genetic maps, and the only way of taking the next step to find out what they actually did would be to switch over to exploit molecular biology techniques. DNA could quite easily be extracted from eukaryotic cells and manipulated in the lab by the end of the 1970s, and thanks to a paper in 1973 by Frank Graham and Alex van der Eb, a method existed for putting DNA, and hence eukaryotic genes, back into animal cells so that their function could be studied in the proper context. It took until 1978 for Gerry Fink and colleagues in the United States, and Jean Beggs in the

United Kingdom to describe the same technique, called *transformation*, for budding yeast, but as soon as he read their papers, Paul realised that the writing was on the wall for fission yeast as a model organism; if there was no analogous transformation procedure, fission yeast biology would die out.

Not everyone was convinced of the potential of molecular biology. Distaste for molecular biology and its purveyors was quite common amongst the more intellectually fastidious geneticists, who viewed it as a rather grubby, menial discipline, where one could get so caught up in the minutiae of getting tricky new methodologies to work that there was no time to think about more important concerns. To some extent, they had a point, because biological problems had a tendency to be parked while techniques were honed to perfection, but those geneticists willing to engage with molecular biology reaped a large reward in terms of data.

The divide is well illustrated in a *Nature* News and Views article that Paul wrote in 1980, at the end of his time in Edinburgh, entitled, "Cell Cycle Control—Both Deterministic and Probabilistic?" The article marks a transition from a theoretical to a real universe in cell cycle theory and points out to the theoreticians that although their work had laid down the important concepts underpinning cell cycle research, it was time to get their hands dirty and start figuring out what was really going on. It was almost a description of Paul's own journey from theory to practice:

> I had all the thinking that those guys had, and all the theory stuff, I was comfortable with that, but I was also very practical. And that was really helpful, because most people who came after that had no idea of the earlier stuff—it sunk without trace. But it was very important for me to make the right research decisions, so it actually influenced me very positively, not because I took it seriously, because I didn't, but because it made me think about the research problems and the issues round them much, much more thoroughly than I would have done otherwise. I owe a lot to it.

Paul had already decided before he left Edinburgh that he was going to concentrate all of his efforts on finding what *cdc2* was doing because something with a crucial role at the two major control points of the cell cycle had to be important. However, once in Sussex, he stopped publishing papers on the cell cycle for a bit and threw all of his meagre resources into the task of getting molecular biology going in fission yeast. Somewhat dramatically, he gave himself a one-year deadline: if at the end of the year he had got nowhere, he would throw in the towel and join the budding yeast community.

There were two aspects to the problem: firstly, the transformation protocol itself, and secondly, getting DNA into a form that the fission yeast transcription machinery would be able to recognise and transcribe into protein. Paul was soon joined by David Beach, an escapee from a nearby lab, and the two of them split the work by category, Paul working on the problem of softening up the yeast cells so that they were ready to take up foreign DNA, and David trying to adapt the plasmid vectors that worked in budding yeast to make them suitable for fission yeast. The two got on well together because Beach was very effective and had a lot of experience of molecular genetics already. Vectors started to appear for testing, but Paul's end of the experiment was not going so well. He had got his technician to make up a set of solutions according to Jean Beggs's protocol for budding yeast, and very unexpectedly, they seemed to work very well for transforming fission yeast too. This was wonderful news, and Paul took the technique with him on a visit to another lab, eager to spread the word. Somewhat embarrassingly, however, when he tried to repeat his success, he couldn't get a single transformed colony and returned home to Brighton with his tail between his legs. In Brighton, transformation continued, mysteriously, to work. For a time it seemed as though there might be a geographical restriction on fission yeast molecular biology, until Paul realised that the crucial solution in the protocol had been made up incorrectly, with one ingredient, sorbitol, missing. Sorbitol turns out to inhibit transformation in fission yeast, so by complete chance, Paul had hit on exactly the right modification to Beggs's recipe. As an example of fortune favouring the slightly incompetent, this story can hardly be bettered, because by honest labour alone, it would almost certainly have taken a really long time to get to the same place.

Having saved himself from the dreadful fate of becoming a budding yeast person, Paul could now get to grips with the problem of cloning fission yeast *cdc* genes, with the highest priority being *cdc2*, the dual regulator of Start and the G_2/M transition. He, Beach, and Barbara Durkacz, a new postdoc in the lab, decided to use a method published in 1980 by Kim Nasmyth, who had turned to the dark side following his PhD with Murdoch Mitchison and was now working on budding yeast as a postdoc in Ben Hall's lab in Seattle.

Nasmyth and Steve Reed, a postdoc with Lee Hartwell, had got together to clone *CDC28*, the key regulator of Start in budding yeast. Their method, which they called complementation cloning, relied on the ability of a normal *CDC28* gene to rescue, or complement, a yeast strain carrying a temperature-sensitive mutant of *CDC28*. Reed and Nasmyth reasoned that normal *CDC28*, although unknown, would be found in a library containing the whole of the budding yeast genome, which they had made by cutting yeast DNA with restriction enzymes and cloning all of the fragments into a plasmid vector. If this library were used to transform a *CDC28* temperature-sensitive strain, the plasmid encoding *CDC28* would be detectable because any colonies carrying it would have a normal cell cycle again and could be picked out by eye from the background of weird-looking cell cycle mutants. This turned out to work very well, and, fortunately, it was a simple matter to adapt the procedure for fission yeast.

Back in Sussex, using a fission yeast genomic library, a plasmid carrying *cdc2* was identified by complementation cloning, but to make the paper a bit cuter, Paul and his colleagues decided to see whether they could redo the experiment, but this time trying to complement mutant *cdc2* using a library from budding yeast. In other words, did *cdc2* have a budding yeast homologue, able to have the same effect on the cell cycle? A budding yeast library was duly screened, and sequences were pulled out that could, indeed, complement the fission yeast *cdc2* mutation. This was a great result because budding yeast and fission yeast are only very distantly related in evolutionary time, having diverged maybe 400 million years ago; finding a gene that worked in both species was pretty amazing. But what was the gene? Because *cdc2* is important at Start, it seemed a good idea to obtain plasmids carrying Start regulatory genes from Steve Reed and test them out individually in the complementation assay. Most surprisingly, the one that worked was Nasmyth and Reed's original *CDC28* plasmid: *cdc2* and *CDC28* were functionally homologous, functionally the same.

The news that *CDC28*, the budding yeast camp's favourite Start gene, was the same as a gene thought predominantly to regulate G_2/M in fission yeast sparked some disbelief to begin with, but the arguments subsided quickly once the truth dawned that both were required in both places. The genetic methods that Lee Hartwell and Paul had originally used to find *CDC28* and *cdc2* had been set up to detect where the genes were most important under conditions of optimum nutrition and rapid cell growth, and the differing lifestyles of the two yeasts meant that in budding yeast, *CDC28* mutants arrested growing cells in G_1, and in fission yeast, *cdc2* mutants caused a G_2 block. Two partial views of cell cycle regulation had merged to create a much clearer picture, thanks to molecular genetics.

Beach, Durkacz, and Nurse published their work in the Christmas 1982 issue of *Nature*, and in the final paragraph, they made a very prescient remark: if two yeasts separated by nearly half a billion years of evolution were using the same mechanism for regulating Start, might it not be possible that the mechanism was so evolutionarily ancient, so entrenched in a cell's life cycle, that control of the analogous Restriction Point in mammalian cells might involve a functional equivalent of the *cdc2* and *CDC28* gene products? In that single sentence lay the seeds of a future Nobel Prize.

Getting fission yeast molecular genetics going had been a real coup for Paul and his small lab, and his rapid reemergence in the pages of *Nature* following his voluntary publishing shortfall should have been a good indicator to the scientific hierarchy that here was a future star in the academic firmament. Paul was enthusiastic, incredibly bright, charismatic, and working on a really important problem, and he should have walked into a job. But again, just as in Edinburgh, he went for many interviews but always came in as the second or third choice. In a profession that should be a meritocracy, he was hampered by the nonscientific prejudices of the hiring committees he faced. He had never worked in the secular cathedrals of genetics or molecular biology, did not have an Oxbridge degree, and did not have sufficiently influential mentors to drop a good word in the right ear. Furthermore, he was not doing the right kind of work, perched as he was on the fence between genetics and molecular biology at a time before the combined weight of fellow converts caused the fence to collapse. In short, he was a risk: he might be as scientifically gifted as he appeared, but he didn't have a thoroughbred lineage, and nobody was prepared to commit themselves.

By 1983, Paul had just one offer, at the European Molecular Biology Laboratory (EMBL) in Heidelberg, but was reluctant to simultaneously ruin his wife's teaching career and uproot her and their two young daughters. Fortunately, he was saved from making the decision. Walter Bodmer, Director of the ICRF, whose eye for scientific talent was in those days unmatched, had noticed that the young upstart in Sussex was lively, ambitious, and scientifically self-confident, and furthermore, was working on a problem, cell cycle control, that lay at the heart of cancer research. The fact that Paul was working on it in yeast was actually a plus, because the human cell cycle research community had all but stalled, and a new approach seemed a good idea. Despite the fact that the ICRF interview committee was less convinced, suggesting to Paul that in addition to his own research, he might also like to run a service facility helping other labs make proteins in fission yeast, Bodmer offered Paul a permanent tenured post, finally giving him the job security he needed.

In 1984, Paul began work in Lincoln's Inn Fields, only slightly worried by hearing an interview with Bodmer on Radio 4 the morning before his first day, in which Bodmer suggested that model systems—flies, worms, frogs, and yeast—might have had their day because so much could now be done in mice and humans. Paul had brought with him the nucleus of an excellent lab, comprising two PhD students, Jacky Hayles and Tony Carr, and two postdocs, Paul Russell and Steve Aves, supplemented with a new technician, Jane Sandall. His leaving Brighton came as a blow for the ladies of the Brighton ICRF charity clothing shop, who were very sad to be losing one of their best customers, but

The Nurse family with Paul Russell (at *left*), ca. 1984.
(Photograph courtesy of Anne Nurse.)

they cheered themselves with the thought that at least they would now be raising money for his work.

At ICRF, the reaction to the newcomers was not entirely positive. Because Paul was the only person working on a nonanimal system in the building and yeast was the most common contaminant of mammalian tissue culture cells, he and his lab were regarded by some as tissue culture poison, shedding spores wherever they walked. There was also disbelief that fission yeast, an organism at least a billion years away from humans, could have any relevance to cancer, something that Paul felt keenly; a *Nature* News and Views he wrote at the time, entitled "Yeast Aids Cancer Research," is a riposte to the doubters, of which there were many, and not just at ICRF. However, Paul is nothing if not stubborn, and he was not about to abandon his long- and passionately held belief that understanding the cell cycle in a simple model would lead to an understanding of it in higher organisms. He just had to provide the evidence, and the time was ripe to do just that; with Paul's own grasp of genetics and molecular biology now supplemented by postdocs and graduate students with expertise in biochemistry and cell biology, the lab was in a fantastic position to capitalise on the long years that Paul had spent characterising mutants, setting up molecular biology, and generally turning fission yeast into a great model organism.

The Paul Nurse of that time is remembered very fondly by his lab. Small, bouncy, and cuddly, with his 1970s-era droopy moustache still firmly in place, he had the energy of a whirlwind and enough charisma to motivate his lab into working hard merely to please him. He juggled the sometimes clashing personalities of the highly ambitious people who had gravitated to him with great skill, at the same time managing to create an atmosphere in which even the newest, lowliest recruit felt able to take part in scientific discussions without risk of humiliation. Sergio Moreno, a Spanish postdoc who arrived on a three-month contract in mid-1986 and stayed for seven years, puts it down to Paul's talent for self-deprecation: "He has this special gift that he behaves in a silly way and then everybody else can do the same—you don't feel bad if you ask something stupid. I think this is very British—in Europe we tend to be more serious!" The group went on regular outings, and they roughed it together in an assortment of grungy youth hostels, where science was fitted in around bracing walks in the rain and complaints about the bad beds and chores. Above all, it was a time of great intellectual intensity and anticipation, of knowing that what was going on in the lab was important.

Paul Nurse and glider, ca. 1985.
(Photograph courtesy of Anne Nurse.)

Iain Hagan, a PhD student with Paul from 1985 onwards, remembers the Nurse lab meetings with particular clarity: "The lab meetings in his group are still one of the highlights of my scientific career. They were amazing—just really, really exciting. Paul wouldn't generally be dominant. He would be letting people discuss things but gently nudging the debate in the right direction." In addition to these scheduled meetings, held for a time in the White Horse pub round the back of ICRF, the lab talked science all the time, chewing problems over for hours in the lab, in the canteen, in the sundry other pubs around Lincoln's Inn, and quite frequently on the number 134 bus from Archway, on which Paul, Sergio, and Jacky came to work.

Driving all of these intense discussions about cell cycle theory, fission yeast lore, and technical troubleshooting was the day-to-day work of the lab, and here Paul's mania for order and tidiness ruled. His own assessment of this is that he is "quite well organised in a shambolic sort of way," but others are more illuminating. Sergio again:

> He's very organised. He can give the impression of being disorganised, but in fact he's someone who is very tidy—have you seen his writing? It's perfect—he's got beautiful writing.

> Everything in the lab was perfectly organised. Every six months, he would ask the technician of the time to find a date, and on that date everybody in the lab, including himself, would wear a lab coat (probably the only time we would wear lab coats!) and then we would clean the lab.
>
> He also likes people who are very tidy and experimentally skilled. He really paid attention to how effective and tidy you were in your experiments, in addition to the intellectual part. Some people weren't as effective as he wanted them to be, and they had to be really smart to compensate for it.

And then, there was Paul's legendary appetite for data, starting with his eagerness for more of it getting in the way of laboratory good manners. Tony Carr remembers that after one key experiment, "I came in in the morning and he'd already been through all the plates, decoded them from my lab books and worked out what the result was. After that I used to hide my lab books on the top shelf where he couldn't reach them." Jacky Hayles, still working with Paul 30 years on, now as joint head of his lab at the London Research Institute, talks about his attention to detail and his ability to see things in experiments that no one else had noticed, although "he can be very irritating when you see it yourself and you want to tell him, but he wants to tell you first."

The next few years of the Nurse lab read like a Greatest Hits album for cell cycle research. Paul, despite his brief flirtation with G_1/S and Start, was still focussed on G_2/M control and what *cdc2* was doing to regulate it, and answers were now coming thick and fast. Viesturs Simanis, a postdoc who arrived in 1985 from David Lane's lab at Imperial College equipped with the ability to get almost any experiment, however difficult, to work, showed that the *cdc2* gene encoded a protein kinase, able to put phosphate groups on other proteins and thereby regulate their activity, and Sergio Moreno went on to show that the kinase activity varied during the cell cycle and peaked just before mitosis, as might be expected for a rate-limiting regulator of G_2/M.

Paul Russell, who had migrated up to London with Paul from Brighton, worked out how the variation in activity occurred by showing that *wee1*, the mutant that Paul had found by accident all those years ago in Edinburgh, was a negative regulator of *cdc2*, and that another cell cycle mutant, *cdc25*, was a positive regulator. Russell had come to Paul's lab from Seattle thanks to a recommendation from Kim Nasmyth, was very focused, and worked fantastically hard, despite having an onerous commute, because he was still living down in Brighton. Tony Carr, another

London–Brighton commuter, recalls competing with him for the coffee-room floor on the nights when neither could make it back home on the last train, although Paul did better than Tony, having equipped himself with a blow-up mattress for sleepovers at the lab.

Russell and Nurse's two *Cell* papers on *cdc25* and *wee1* in 1986 and 1987, and Simanis and Nurse's on *cdc2* in 1986, together with the constellation of papers surrounding them, were the vivid proofs of the theories that Paul had formulated back in Norwich as a PhD student. In the intervening years, Paul, with the help of his colleagues, had worked out how to find rate-limiting steps in the cell cycle, had gone out and isolated mutants in those steps, had taught a whole field how to do molecular genetics, and now knew the DNA sequences of the mutants and could see exactly how they worked together. The whole saga was an amazing tour de force, especially given the stunning lack of interest shown by the rest of the scientific world. *cdc2*, the master kinase that was responsible for flipping the on/off switch for the G_2/M transition, was itself decorated with phosphate groups (phosphorylated) by the *wee1* proteins, also kinases, and this phosphate decoration inactivated it. When *cdc2* needed to be switched on, *cdc25*, which encoded a phosphatase (a protein able to remove phosphate groups), undid *wee1*'s work, and *cdc2* protein became active. It was beautifully simple. But did it matter anywhere other than yeast? Paul, pretty much alone in the field as usual, thought that it did.

Paul's conviction that *cdc2* would be present in other species besides yeast had been bubbling under for some time, because he knew that entry into mitosis from G_2 was regulated just as tightly in higher eukaryotes as it was in fission yeast. In 1971, Yoshio Masui, working on frog oocytes, had discovered the existence of a mysterious activity, which he called Maturation Promoting Factor, or MPF. Frog oocytes arrest naturally in G_2 before fertilisation but can be forced to mature and go through mitosis by injection of MPF, contained in an extract made from mature egg cytoplasm. MPF activity turned out to be present in all of the higher eukaryotic cell types, including mammalian, that could be tested, and to be fundamentally important; it was a rate-limiting G_2/M regulator just like *cdc2*, but its true identity was still unknown by the early 1980s, despite strenuous attempts to purify it from the cytoplasmic soup in which it hid.

While he was in Sussex, Paul had started a collaboration with Chris Ford, a frog person there, to see whether injection of *cdc2* into oocytes could induce maturation, but despite some tantalising hints that something might be going on, the data were too inconsistent to be convincing, and the experiment had been abandoned. *cdc2* might well have something to do with MPF, but clearly, another way had to be found to test the theory.

In 1985, Paul hired a new postdoc from Jean Beggs's lab at Imperial College. Melanie Lee was looking for a second postdoc, and knowing that Viesturs Simanis, an old friend from Imperial, was having a great time in the Nurse lab, decided to apply there herself. The application process started in a pub near ICRF, where Melanie discovered Paul, "scruffy, little and very vibrant," having a drink with a vision of male pulchritude "all in white, tall and Adonis-like," who turned out to be Kim Nasmyth. Paul, much to her astonishment, said yes to Melanie almost immediately, and she started work shortly afterwards. Her first day in the lab featured a meeting with the boss to discuss possible projects, and she remembers Paul presenting her with two options: "He said, 'You can do one of two projects. You can work on *cdc10*, which is fairly ordinary, or you can work on a project that's high risk but justifies my presence in a cancer institute; that is, you clone human *cdc2*.'" Melanie, without hesitation, went for *cdc2*.

Paul had tried to sell the human *cdc2* idea to other new lab entrants but had got no takers, because of the high likelihood that it would be a total bust. Of course, by Paul's lights, this made it a fine project: "If I wasn't doing things where I thought there was a reasonable chance of failure, I wouldn't think we were doing anything important. I begin to realise I'm a bit odd in that way." Spending years looking for a protein that might exist only in the mind of their boss was not a high priority for anyone hoping to get a good job on the back of their postdoc work. Besides, there was so much else to be done that it was easy to pick less-risky winners; other projects might require elegant genetics and tricky biochemistry, but as long as you were good at the bench and worked hard, the payoffs were almost guaranteed.

Melanie Lee and Freeway, 1987.
(Photograph courtesy of Melanie Lee.)

Melanie, however, was a bit different. This was her second postdoc, so she wanted to do something startlingly good. More importantly, she was a risk taker, just like her new boss, and the thought of doing something quite so wacky was exciting; if it was really possible that *cdc2* had jumped the billion years or so of evolution dividing humans from fission yeast, she wanted to be the person to find out.

There was another factor too, that of her relative inexperience in the field. Unlike the rest of the lab, she came from a non-fission-yeast, non-cell-cycle background, and she lacked the knowledge to dispute Paul's judgement that the project had a chance. As Paul says: "Melanie was the only one who said she'd have a go at it, because I said in my view this was the most interesting project in the lab. She was receptive to my advice, which was useful for a project like hers. I was suggesting she did quite bold things and she was prepared to trust me. I don't think the others would have necessarily done that.... She was also extremely capable and stable, and really efficient and effective, which was necessary for this project."

The human *cdc2* project illustrates very well the hard slog and serial failures that accompany most scientific ventures. After many months of painfully slow nonprogress, it became very obvious to Melanie why Paul had emphasised the difficulty of the project and why nobody else had wanted to take it on. Very simply, there appeared to be no good way of even looking for a human *cdc2* homologue by conventional methods, let alone finding something.

When searching for relatives of known proteins, there were at that time two standard ways to proceed: look for similarities at the DNA level and find the gene encoding the protein, or search for similarities at the protein level and then work back to the gene. Both methods were molecular fishing expeditions, in Melanie's case using fission yeast baits dipped into enormous pools of human genes or proteins. To search by DNA homology, Melanie probed a human gene library with an antisense version of the fission yeast *cdc2* gene sequence, hoping that the fission yeast antisense DNA would be able to pick out its human sense counterpart. To look for protein homology, she used an antibody able to recognise fission yeast *cdc2*, this time fishing in a human protein library, hoping that the antibody would recognise some shared structural component enabling it to hook out human *cdc2*.

The workload involved in spreading the huge libraries on to hundreds of soup-bowl-sized agar plates, probing them, and then processing the results was enormous, and after a few months, Paul gave Melanie a technician, Martin Goss. However, despite receiving a great deal of help from Viesturs Simanis, who had made the anti-*cdc2* antibodies and was a whiz with the type of hybridisation experiments that Melanie and Martin had to do, they got precisely nowhere. Either the baits pulled out nothing, or under less stringent conditions, they pulled out junk. Too much evolutionary time had elapsed between fission yeast and humans, and if there was indeed a human *cdc2* gene, it was untouched by the molecular overtures from its distant relative.

Matters came to a head at a group meeting in the summer of 1986, about nine months after Melanie had started, when as usual, she had to present yet another litany of failure to the assembled lab. It was obvious that pursuing her current strategy was pointless, but what else was there? The only other method of finding homology was by functionality, as Beach, Nurse, and Durkacz had done when they had discovered that *cdc2* and *CDC28* were doing the same thing in fission and budding yeast, but how on earth could one do that between yeast and human? Leaving aside the sheer unlikeliness of a human gene being close enough in function to be able to rescue a fission yeast mutant, there was no way of doing the experiment, unless there was a human protein library that would work in fission yeast, which, surely, there wasn't.

And then someone—Tony Carr, Viesturs Simanis, nobody can quite remember who—realised that there was. It was probably only the Nurse lab that had the combined knowledge and audacity to contemplate it, but there *was* a library that might, just might, work. It had been made by Hiroto Okayama in Paul Berg's lab in Stanford, California, and funnily enough, Melanie had already been using it for her fruitless fishing experiments. The Okayama and Berg library contained pretty much every gene in the human genome cloned into a plasmid vector able to grow in *Escherichia coli* bacteria, just like many standard libraries from that time. However, the neat factor about it was that Okayama had worked out how to get the library to switch on, or *express*, all of its proteins in mammalian cells as well as *E. coli*, by topping and tailing the gene sequences with control elements that the fussy transcriptional machinery of mammalian cells could recognise and use.

The reason that nobody had thought of using the Okayama and Berg library for a complementation experiment in fission yeast before was that just as mammalian cells are very picky about their transcriptional control elements, so are some yeasts. Budding yeast was well known only to express proteins using its native control elements, and everyone had assumed that fission yeast was equally xenophobic. However, Paul's lab had recently been tinkering with the particular control element, called the SV40 early promoter, that Okayama had used to drive his library genes, and they had found to their surprise that it worked quite happily in fission yeast, which turned out to be far less fussy than its budding yeast cousin. There was a slim chance that the Okayama and Berg library could be coaxed into expressing human proteins in fission yeast cells.

Even if the library could be put into fission yeast, many further obstacles loomed, but as the lab discussed the idea with a mounting sense of excitement, the realisation flew around the room that although it would be bold to try it, it was not totally foolhardy. One by one, the requirements for getting the experiment on the road fell into place. There were two main issues. First was the question of what strain of yeast the library should be put into. Paul and Jacky had that one nailed instantly, because they knew that they had a *cdc2* temperature-sensitive mutant strain, *cdc2-33 leu1-32*, which was totally watertight, stopping absolutely dead at the restrictive temperature. If the library went into that and any colonies grew at the higher temperature, then it had to be caused by an incoming gene rescuing the defective *cdc2*. Next was how to tell whether the library had made it into the yeast cells at all—the transformation procedure had to be maximally efficient to make sure that every human gene possible was represented. The way round that one was to use the Okayama and Berg plasmids to piggyback a second plasmid carrying a selectable marker, *LEU2*, into the cells. If the transformed cells were grown in medium lacking leucine, only those carrying the *LEU2* plasmid would survive, because the *cdc2* mutant strain that Paul and Jacky had suggested using had been modified so that it could not synthesise leucine on its own. The bare bones of the experiment were there: transform the library into *cdc2-33 leu1-32* cells, along with the *LEU2* plasmid, and plate at the permissive temperature in the absence of leucine; after a day, when any transformed cells would have had a chance to recover and start growing, shift to the restrictive temperature, and look for any colonies that could grow normally. Easy.

Nobody thought that it would work. Peter Goodfellow, whose lab was next door, came by Melanie's bench almost every day, and invariably said, "stupid experiment—it'll never work." Even Paul was a bit dubious. It was four years since he had published the *cdc2* and *CDC28* complementation paper, the human project had been on his mind ever since, and he had had ample time to try to make a library himself, but he "wasn't prepared to ... because it was such a long shot." It was an act of desperation, but the one fact in its favour was that it was a quick experiment, and Melanie and Martin would know soon enough, one way or the other.

Like decorating, the secret of doing successful experiments lies in the time spent on preparation, and this experiment took a lot of forethought. The logistics were frightening. Normally, yeast transformations are done using about 10 million cells and ~1 μg (a millionth of a gram) of plasmid DNA. To get enough colonies to ensure that every human gene had a decent chance of getting into a yeast cell, the procedure had to be amplified a hundred-fold; unfortunately, this was not just a matter of finding a big bucket to put all the cells in, but of doing a hundred separate transformations, because the protocol did not scale up efficiently. The hundred transformations then had to be spread onto hundreds of agar plates, which all had to be jammed into incubators, and then the massed plates had to be checked daily for growing colonies. It was just as well that Melanie was, in Paul's words "about the most well-organized worker I've ever had," and that Martin Goss, in Melanie's opinion, was "perfect—he didn't care whether it was a mad experiment or not, he still provided the technical help, and he was a good pair of hands at the bench, a jolly good technician."

After a couple of small-scale runs, to maximise the transformation efficiency, Melanie and Martin ran their first big experiment, got nothing, and managed to really annoy the rest of the lab all at the same time; putting a vast number of coldish agar plates into the lab's 36°C incubators turned out to cause such a drastic drop in temperature that everyone else's temperature-sensitive mutants started growing again, wrecking a good few experiments. They tried again, being a bit more careful about the temperature issue. After two weeks, "just when you're about to throw the plates out because you're fed up with it," they saw one or two rather pathetic colonies struggling along on the otherwise empty plates. By this time, Melanie and Martin were on a roll, so they set up another huge experiment and left the plates to cook whilst Melanie flew off to Banff in Alberta for that year's annual yeast scientific beano, more formally known as the 13th International Conference on Yeast Genetics.

In Banff, Melanie had an interesting time fending off David Beach, who had learnt on the grapevine that the Nurse lab were looking for a human *cdc2* homologue. Beach, now running his own operation at

Cold Spring Harbor, had also been trying to clone human *cdc2*, had got nowhere, and was desperate to find out what was going on in London. Relations between him and Paul having deteriorated since their days together in Sussex, he figured that perhaps the postdoc doing the work would be more forthcoming and cornered her one night in the bar. A rather surprised Melanie fought off the barrage of questions valiantly and managed to escape in the end, with Beach none the wiser, but considerably crosser. In a way, it was quite reassuring to know that someone else in the world thought the project was worth fighting over, but the experience was less than pleasant.

Back in London after the meeting, any vestiges of jet lag and alcohol poisoning dissipated instantly when Melanie looked at the latest experiment. This time, there were five colonies, and they looked really promising. There was an emergency lab meeting to decide what to do. Because fission yeast is very good at scrambling introduced DNA sequences, sometimes disassembling plasmids and stuffing them into its own genome, the most urgent task was to rescue the plasmid DNA from the cells before it vanished. The colonies were plated out on new agar to amplify them up, and when Paul came to take a look at them, he realised that they were really onto something, because whatever was rescuing the large *cdc2* mutant cells had to have come from a library plasmid. Sergio Moreno, who had arrived in the lab only a few weeks previously, remembers: "Paul was not doing experiments at the time, all he was doing then was looking at the cells—he really liked to look at cells under the microscope. It was clear when he looked at [one of] the colonies that it was really growing. When he picked the colony and spread it he realised there were small cells that were the complementing ones mixed with large cells that were losing the plasmid. He was really excited about that and everybody in the lab was aware this was going to be very important."

In the event, only two of the five colonies still had rescuable plasmid DNA in them, but it was enough. Melanie transformed the precious DNA back into the *cdc2-33 leu1-32* cells, to see whether either plasmid would complement the *cdc2* mutation, and to wild excitement, she hit the jackpot with both; at the restrictive temperature, *cdc2-33 leu1-32* cells were able to cycle normally as long as the plasmids were present. Comparison of the two plasmids showed that they contained an identical DNA insert, and the insert was also present, although chewed up, in the three other original colonies from the screen.

Being at the frontiers of science carries with it a certain degree of fear; it is frightening to be out of the limits, chasing after things that might not exist, and good scientists live with varying levels of anxiety that any new work they do might be wrong. And so it was in the Nurse lab. Finding human *cdc2*, particularly by such a frankly bonkers approach, was so important and so unlikely that wholly justifiable paranoia started to set in. What if the inserts were contaminants, that somehow a yeast *cdc2* or *CDC28* had sneaked into the library, or had just been in the air in the lab and got into the hundreds of plates undetected? The possibility of the result being an artefact was alarmingly likely, and the only way of finding out was to establish the DNA sequence of the inserts, which was going to take a few weeks. Melanie and Martin started sequencing, and the lab held its collective breath.

Paul was probably most affected:

> I didn't want to burn us, by just following nonsense. When the clones came up, when we showed it was due to a plasmid ... what I remember is that every time Melanie did something, we discussed why it might not be saying what we hoped it would say. I then went through this psychologically peculiar time where I assumed it would fail at each step, but we just kept going. I was in this schizoid thing—I was keeping myself safe assuming it wouldn't work, then I'd go home at night and think, "I'm just going to imagine it has worked, because I'll go in tomorrow and it will be gone."

Sequencing of the 2000 bp (base pairs) of DNA in the inserts carried on throughout the autumn, but by Christmas 1986, the work was finished and the complete sequence assembled. The good news was that the DNA sequence didn't match either the fission yeast or the budding yeast *cdc2* genes, and thus had to be something different, but depressingly, there were no obvious similarities between the new sequence and the yeast genes at the DNA level. But what about at the protein level? The yeast *cdc2* and *CDC28* proteins, although not completely similar, have one region that is invariant because it is needed for their activity as protein kinases. The region has the amino acid sequence proline–serine–threonine–alanine–isoleucine–arginine–glutamate, commonly abbreviated using the amino acid one-letter code as PSTAIRE. Melanie ran her new sequence through a programme that could decode the DNA, translating it to see what amino acids any protein made from it would contain, and with literally bated breath, she, Paul, and the lab stood over the old dot matrix printer as it churned out the result.

Over to Melanie: "I will never forget that day. The printout came off with PSTAIRE on it. It was unbelievable. We were all watching it and I said, 'There *is* PSTAIRE there!' and Paul said, 'Are you sure?' I said, 'Yes, but it's a different DNA sequence, it's completely different.' I remember Paul Russell said, 'It's a Eureka moment! It's worked, it's amazing!' Paul was very excited and I was like, 'Oof, thank goodness for that!'"

After that wonderful, climactic moment, there was still a fair amount of work to be done to show that the new sequence really did originate from a human source and to show that human cdc2 protein could be detected by the yeast cdc2 antibody (easier to do now that it was obvious what they were looking for), but the paper by Lee and Nurse, "Complementation Used to Clone a Human Homologue of the Fission Yeast Cell Cycle Control Gene *cdc2*," was submitted to *Nature* in March and came out in early May 1987. The last sentence of the article was a vindication of Paul's vision: "The identification of a *cdc2*-like function in human cells suggests that elements of the mechanism by which the cell cycle is controlled will probably be found in all eukaryotic cells."

The appearance of such a major paper, the culmination of many years of speculation, hard work, boldness, and the odd bit of luck, should have been an occasion of great celebration for Paul and Melanie, but unfortunately, things didn't quite turn out that way. The paper came out while both its authors were attending a conference in Heidelberg, and Paul was due to give a talk about its contents towards the end of the meeting. However, just before leaving Britain, Paul had picked up an ear infection, the pain of which was exacerbated by the flight from London to Frankfurt, and he ended up being admitted to the local hospital, spending most of the meeting there. He did manage to show up for his talk, but the import of what he was saying was overshadowed by his head being swathed in a huge white bandage. "I was in such pain I can't even remember what I talked about. I had a horrible ear infection—it was unbelievably painful. The hospital had to puncture my eardrum and drain it. It was awful, awful. The worst pain." Melanie fared little better, but for a nicer reason. She was in the first trimester of her first pregnancy and had awful morning sickness. Her summary of their triumph says it all: "I felt dreadful and Paul was really poorly."

Paul and Melanie's paper started a revolution in cell cycle research. Because all eukaryotes turned out to have *cdc2*-like genes, the genetics established in fission and budding yeast laid down a roadmap of what the biochemists in higher eukaryotes should be looking for. Research leapt ahead when the two fields synergised, and Paul suggests that the yeast work may have saved somewhere between five and 10 years of research in mammalian systems.

One of the first and most important molecules to fall to the newly unified field was MPF, Maturation Promoting Factor. In 1988, after almost two decades of effort, MPF was purified by Jim Maller's lab in Denver and was shown to consist of two proteins, one of which was a kinase. Using antibodies raised against the PSTAIRE region of *cdc2*, two collaborative teams, the Nurse and Maller labs, and David Beach and John Newport's labs, published back-to-back papers in *Cell* showing that the MPF kinase was, indeed, *cdc2*, as Paul had almost shown with Chris Ford back in Sussex in 1980. The primary mechanism regulating G_2/M in both yeast and higher eukaryotes was not just similar; it was identical.

In a particularly nice twist, MPF was also the means of uniting the work of two researchers who met in the early 1980s, hit it off instantly, and have kept the cell cycle world entertained with their energetic sparring ever since. When discussing science, Tim Hunt and Paul Nurse have the air of two small boys playing on a beach together, falling out, and making up, endlessly fascinated by the rock pools, the funny looking shells, and the task of building huge and sometimes precarious sandcastles. In 1982, Tim had accidentally stumbled into the cell cycle world through his work on protein degradation in sea urchin eggs, when he discovered cyclins, the keys to how the cell cycle turns. Cyclins must be made, degraded, and then made anew in order for cells to divide properly, and the reason for this became clear once MPF was purified. MPF comprises one molecule of Tim's cyclin protein bound to one molecule of Paul's cdc2 protein, and cdc2 cannot work without the assistance of cyclin; indeed, cdc2's official biological name, CDK2, cyclin-dependent kinase 2, shows the defining nature of the interaction. Functional MPF is created when enough cyclin has built up to switch on cdc2 activity, thereby driving cells into mitosis. Cells cannot exit mitosis unless MPF activity is switched off, and for this to happen, cyclin must be degraded. At the end of mitosis, the cell divides, cyclin starts to build up again, and the process is repeated.

Tim's cyclin and Paul's cdc2 were the founder members of two protein families, each with multiple related members. Today we know that progression through the cell cycle in higher eukaryotes is driven by a cell cycle engine composed of particular CDKs in partnership with different cyclins. Distinct cyclin–CDK pairs are needed at different points in the cycle, but they all work in the same way, switched on by the presence of cyclin in order to push cells through a rate-limiting control step, and then switched off by cyclin degradation when the step has been passed (a good example of this is John Diffley's preRC, one of the stars of Chapter 4). By means of phosphorylation and dephosphorylation by enzymes such as wee1 and cdc25, and the action of CDK inhibitors, the cell cycle engine is tuned to respond with exquisite sensitivity to changes in the environment of the cell in which it resides, going faster or more slowly, or stopping altogether depending on the prevailing conditions. As might be expected, the complexity of this regulation is greater in multicellular organisms such as ourselves than in unicellular yeasts, but the basic principles of cell cycle control laid down by Hartwell and Nurse have held firm. Not surprisingly, the two geneticists have won any number of prizes over the years, culminating in the award of the 2001 Nobel Prize in Physiology or Medicine, which they shared with Tim Hunt.

Nowadays, despite being loaded down with awards, a knighthood, and far too much administration in his current schizophrenic incarnation as President of the Royal Society and CEO of the Crick Institute, Paul's heart still belongs very much to the lab; seeing him discussing

The Hug: Paul Nurse and Tim Hunt, 2001 Nobel Prize ceremony, Stockholm.
(Photograph courtesy of epa/Gerry Penny.)

science with his group reveals a happier, more relaxed persona, and he has never lost the early idealism that led him into research:

> I think it's a privilege to do what we do. I did make a pact with myself. I thought when I did a PhD, "Should I do something obviously useful, like work on malaria, for example, where contributing to our understanding, even in a small way, might be useful, or could I just indulge my own curiosity and work on whatever I like? And what I decided was, as long as I'm at the top of the tree I can do the latter. But if I slip from that then I will shift onto something more obviously useful." And you'll notice that I've had administrative, managerial posts for a long long time. People don't believe me, but I don't actually really enjoy it and I'm not even that good at it. I never remember half the things I'm supposed to remember or do. I do it because I feel I'm paying back a debt. So by spending half my time doing this other stuff, I justify what I do in the lab.

The last word goes to Melanie. After an extremely successful career in the pharmaceutical and biotech sector, she looks back on her adventures with *cdc2* thus: "Martin Goss played a big role—all the lab played a big role. Paul was utterly convinced it was there, and the lab, who had more experience than me of fission yeast, were the ones saying, do these wild and wacky things, especially Tony Carr and Viesturs. I didn't detect any jealousy from anybody—they were all lovely. It surprised me that this project was available, but it was available because it couldn't

Nurse lab outing, ca. 1987.
(Photograph courtesy of Sergio Moreno.)

be done. The fact that Paul and I came together for those two years was meant to happen."

Web Resources

www.bbc.co.uk/programmes/p0094839 Paul's Desert Island Discs.

www.nobelprize.org/nobel_prizes/medicine/laureates/2001/nurse-autobio.html# Paul Nurse, Tim Hunt, and Lee Hartwell on the Nobel Prize website.

Further Reading

Murray A, Hunt T. 1993. *The cell cycle: An introduction.* Freeman, New York. The clearest and most interesting book I found about the cell cycle amongst the many out there.

CHAPTER 7

Death and Glory

The tissue culture facilities of a modern biological science laboratory can easily be distinguished from the rest of the lab by their extreme cleanliness and, relatively speaking, tidiness. Notices abound, exhorting users to put stuff away properly, chuck out old bottles of solutions, soak glassware in the correct buckets, and dispose of plastics in the special bins labeled "autoclave waste." All of these measures, unusual in the cheerful anarchy of most labs, are designed to ward off the dreaded twin spectres haunting all tissue culture work: that the cells die because they haven't been looked after properly, or that somebody (never the cells' owner, naturally) contaminates the precious cultures with bacteria or fungus. Such calamities at best cause a few days' labour to be consigned to the bin, but sometimes, months of painstaking work are destroyed in a few moments of carelessly insanitary behaviour. Have you been making bread at the weekend? Then don't go near your cells, or your baking yeast will mysteriously manage to breach all of the defensive barriers and get into your cultures.

As befits a black art, tissue culture is surrounded by rituals evolved to ward off death and disaster. Doing tissue culture requires a mixture of skill, concentration, and willingness to spend extremely long hours doing mindless repetitive tasks. Upon entering the tissue culture lab, the modern acolyte puts on a specially designated lab coat and luridly coloured latex gloves, and gets out the chilled cell growth medium from the fridge to warm up in a water bath set to blood temperature. The names of the tissue culture media contain their history: Dulbecco's Modified Eagle's Medium; Roswell Park Memorial Institute Medium; Ham's Nutrient Mixture—all were developed by trial and much error and contain a witches' brew of chemicals that, when mixed with serum derived from the blood of cows, chickens, and other farm animals, will nurture the precious cells. Bottles of medium are jealously guarded, sometimes to the point of being labeled in a paranoid personal code, and are never used for multiple cell types, to prevent cross-contamination. Woe betide the lab worker caught infringing this rule by their boss, who has a folk memory of the terrible period when practically every cell line turned out to have been contaminated with the weed-like HeLa cells.

The growing cells are taken from their temperature-controlled incubator, bringing with them the tantalisingly fizzy whiff of CO_2-laden air, and looked at under the microscope, to see whether they are "ready to split"; whether they are starting to outgrow their flask. If they are, one of the big tissue culture hoods is fired up. The air flow is switched on, and the fans rumble into life, providing a barrier of constantly rushing air between the user and the enclosed work surface. Then the front panel is removed, and the inside of the hood is thoroughly swabbed with alcohol, together with all of the bottles of medium and miscellaneous bits of equipment. In former days, enthusiastic workers would set the alcohol on fire "just to make sure" that everything was truly sterile, although the side effect of sometimes setting fire to one's alcohol-soaked hands (gloves are a modern innovation) in a spectacular but (mostly) painless way may have been a more likely enticement.

The work begins. Endless flasks are brought to the hood, and the cells inside are laboriously removed and counted under the microscope. Then, they are divided up into new flasks pre-filled with growth medium and returned to the warmth and darkness of their incubators. Once the mundane husbandry is finished, there are likely to be experiments to tend to as well. Reagents are added to cells, medium is changed or augmented, and cells are harvested and go off to be analysed back in the real world of the main lab. Hours and hours of time can be spent in tissue culture, often to the accompaniment of appalling rock music, mercifully piped straight into the ear nowadays, but formerly blasted out on boom boxes to willing and unwilling listeners. Conflicts between the Radio 3 cricket commentary brigade and the lovers of The Smiths and the Stones were legion before Mr. Sony came to their rescue. And in the background, the fans driving the airflow of the hoods provide a steady rumbling ostinato.

Despite all of the precautions and the hours spent cosseting the cells, everyone who has spent any time at all doing tissue culture knows that the one given is that cells die; some quickly, some slowly, some by infection, some by rough treatment, some by voodoo, but they die.

Paradoxically, cancer cells are generally much better at dying off in tissue culture than fresh cells derived from normal tissue. Dying cells are a nuisance, because you can't do any experiments with dead cells, and they are a mark of scientific incompetence—you must have forgotten something and are therefore a science numpty.

Why this lengthy preamble? Because, in the face of all of the cumulative tissue culture lore, the mountains of dead cells, and the wasted years of scientific time, in one laboratory at the Imperial Cancer Research Fund, someone finally started to wonder whether all of this death might be trying to tell us something biologically meaningful. Was it possible that cell death was not just an awkward obstacle to doing decent experiments, but might matter? Furthermore, might finding the reason for all of the cell death lead to the answer to an enduring problem with our understanding of cancer: why is it that cancer, a disease of madly proliferating cells, is not more common?

Today, Gerard Evan is a major force in cancer research, but in the mid-1980s, he was a distinctly worried junior faculty member in the Ludwig Institute of Cancer Research in Cambridge. Gerard's career up to then had had the standard upward trajectory of a smart Oxbridge-trained scientist: a PhD at the Medical Research Council (MRC) Laboratory of Molecular Biology in Cambridge, followed by a very successful postdoc in future Nobel laureate Mike Bishop's lab in San Francisco, and then a return to Cambridge in the early 1980s to run his own lab as a tenure-track group leader, the lowest rung on the lab head career ladder.

Gerard is a very appealing character; a curly-headed, eminently cuddly man, his quicksilver mind is overlaid with a mixture of charm and irrepressible enthusiasm, rendered bearable for those of a more Eeyore-ish disposition by an undercurrent of self-doubt and angst. His hunger for facts of any kind means that his brain is stuffed with information, both trivial and less so, which makes him an entertaining companion despite his appalling taste in jokes. He is ridiculously easy to tease; friends have a wide choice of strange enthusiasms to pick from: a shady past as a member of the Comberton Morris Men; an unwholesome addiction to Rohan multi-zip trousers; his

Gerard Evan picnicking in France, ca. 1990. *(Photograph courtesy of author.)*

unswerving dedication to being an early adopter of any and every piece of electronic gadgetry—the list is endless. Scientifically, he possesses in spades the gift of being able to integrate information from diverse sources and apply it to his own work, giving him the ability to view data from multiple perspectives; he always sees the wood despite the trees, an invaluable asset when hacking one's way through the murky data forests encountered in biology.

Gerard had spent his postdoc working on an oncogene called *MYC* (pronounced Mick), identified in the Bishop lab in 1978 as the gene in the chicken retrovirus MC29 responsible for its cancer-causing activities. v-*myc* was soon shown to have a cellular counterpart, c-*MYC* (known as *MYC* hereafter), which could be converted into an oncogene when over-expressed. By 1984, when Gerard returned to the United Kingdom, MYC was a well-established oncoprotein and had been shown to be able to drive uncontrolled growth when introduced into cells in tissue culture. (A quick note about gene nomenclature: names in italics signify genes, names in normal font signify their protein products. Hence, the *MYC* gene encodes the MYC protein. Strictly speaking, there is another level of hierarchy depending on what species the gene in question comes from, but I am ignoring that here for reasons of clarity.)

Like many nascent lab heads, Gerard had a dowry from his ex-boss comprising unfinished postdoctoral projects, and he simply continued to work on them. He was allotted a minute amount of lab space in the Ludwig Institute and hired a young research assistant, Dave Hancock, who'd previously been working across the road in the Department of Pathology and Virology but fancied a change. Dave was soon joined part-time by Trevor Littlewood, who was actually the laboratory manager for the Ludwig Institute but was pining for the bench; in fact, he moved into a full-time position with Gerard shortly afterwards, leaving the world of laboratory administration behind forever.

Trevor and Dave are pivotal to the story of Gerard Evan. Both very good at lab work, both good humoured, tolerant, well adjusted, and equipped with very English senses of humour, they provided a foil for their hyperactive boss; nobody could have asked for a better start to their lab than having Trevor and Dave around. The three of them produced some sound but not particularly inspiring work on *MYC*, exploiting the monoclonal antibodies that Gerard had made in the Bishop lab. The recent development of monoclonals meant that there was a lot of low-hanging experimental fruit to be picked for those in the know;

Dave Hancock in the laboratory, 1980s.
(Photograph courtesy of Trevor Littlewood.)

monoclonals were extremely precise tools for tracking where in the cell a protein was and whether it was appearing and disappearing in some coordinated way or staying around for the whole of a cell's life. Early on, in 1985, Dave and Gerard hit the high spot by publishing a paper in *Cell* showing where in the nucleus of a cell the MYC protein lived.

The lab's publication output over the next couple of years was good, but not mind-blowingly so, and Gerard's growing feeling that, scientifically, he was drifting without purpose, was compounded by his disappointment with the Ludwig Institute, which was not proving to be the vibrant and well-connected place that he had hoped it would be. In 1987, therefore, he started looking around elsewhere for jobs. The ICRF was advertising for junior tenure-track group leaders, the same career rung that Gerard was currently on, and he applied for and was offered a job there. However, his bosses at the Ludwig Institute conducted a charm offensive to persuade him to stay put, and in the end he turned down the offer from London.

In retrospect, it is somewhat incomprehensible that the Ludwig wanted to keep him, because shortly afterwards, they announced the closure of their Cambridge laboratories, with the concomitant redundancy of all its employees. This was a calamity for many of those there and a period

Dave Hancock and Trevor Littlewood, Evan laboratory dinner, 1980s. *(Photograph courtesy of Trevor Littlewood.)*

in his life that Gerard still recalls with much pain. After the initial shock had subsided, the Evan lab decided that they should if at all possible ride out the storm together, trying to move en masse elsewhere, and Gerard called Walter Bodmer, the ICRF Director, to learn whether there were any jobs still on offer. Walter, whose sometimes alarming demeanour is widely held to conceal a kind heart, not to mention an impeccably good nose for scientific excellence, not only encouraged Gerard to move, but offered him a permanent tenured position at Dominion House, an ICRF outpost over by Bart's Hospital. Despite reservations about the lengthy (more than two-hour) commute that would be required (family constraints meant that none of the Evan lab members, including Gerard, could relocate to London), Gerard accepted the job, and in 1988, the lab moved.

The Dominion House laboratories had opened in 1983, as part of Bodmer's drive to expand the ICRF's links with clinically relevant cancer research. Under the initial leadership of John Wyke, they were small, comprising just four labs, and from the outset, suffered from being a rather isolated satellite of the main laboratories in Lincoln's Inn Fields, without the focus that was to make Clare Hall so successful. By 1988, John Wyke had moved on to become Director of the Beatson Institute in Glasgow, and the Dominion House lab heads comprised Mike Owen, an immunologist, and Gordon Peters and Roger Watson, both interested in the molecular biology of oncogenes. Life there was a bit quiet, with occasional excursions to Lincoln's Inn Fields, a half-hour walk away, being the only excitement; thus, the arrival of the Evan lab was a major social and scientific improvement. Because science is rather incestuous, Gerard and his three new colleagues already knew and liked each other pretty well. Having Gerard bouncing about the corridors galvanised the place, and the Evan laboratory also had a sense of a new beginning.

Gerard's work up to then had been focused on the biochemistry of MYC: finding out where the protein was located, when it was made, when it was degraded—the nuts and bolts of a protein's life. However, the move down to London precipitated a change in focus, towards a far more biological emphasis. It was all very well knowing when and where a protein lived and worked, so to speak, but that told one very little about what it did, which in the end is what matters in biology. The scientific environment at the ICRF was fertile ground for Gerard's shift in thinking; as well as his new Dominion House colleagues, he had a close scientific friend in the ICRF main laboratories: Hucky (Hartmut) Land, an eccentric bear of a man with an uncanny resemblance to an extremely large garden gnome. Hucky had done a postdoc in Boston with the rather shorter Bob Weinberg, who, somewhat paradoxically given his size, is one of the giants of the oncogene field. Together with Luis Parada, Land and Weinberg, as related by Bob Kamen in **Chapter 2**, had published an extremely important paper in *Nature* in 1983, showing that for the oncogene *RAS* to turn normal cells into tumour cells, it required a second oncogene to help it; in other words, oncogenes had to cooperate with each other to cause cancer. Parada, Land, and Weinberg suggested that each of the oncogenic partners brought a separate oncogenic attribute to the relationship, and only when these were added together could the normal cellular controls be overridden, allowing the formation of a cancer. Hucky had been recruited as an ICRF lab head in 1985, and his lab had continued to work on oncogene cooperation, with a strong interest in what MYC and RAS were able to do together.

All of the Land lab's work was based on tissue culture, and in order to get the genes in which they were interested into the cells in which they wanted to study them, they had developed a range of retroviral vectors, the pBABE series, which were so useful that they are still widely available today, 25 years later. To put a gene into a cell, one inserted its DNA into a suitable pBABE vector, packaged the naked vector DNA up in a protein and lipid coat to make it look like a retrovirus, and then infected the cells of interest, selecting the ones that had taken up the pBABE invader by treating them with an antibiotic. The protein antidote to the antibiotic was encoded by a gene in the pBABE vector DNA, so that cells without pBABE died from the antibiotic treatment,

Hucky Land, 2002. *(Photograph courtesy of CRUK London Research Institute Archives.)*

and those carrying pBABE lived and grew, becoming so-called stable cell lines permanently expressing the foreign gene of interest. pBABE vectors made it possible to make hundreds of stable cell lines, and some of those in Hucky's lab contained an introduced *MYC* oncogene, which piqued Gerard's interest.

The world of molecular biology at that time was still in a state of great excitement about the cell cycle, following Lee and Nurse's 1987 publication showing that the cell cycle control mechanisms were pretty much universal in eukaryotic organisms. Legions of postdocs and graduate students were occupied worldwide in working out how their protein of interest could be bolted onto the cell cycle: Did it regulate the cycle? Did it help cells get into cycle? Did it stop them getting out of cycle? Did cells cycle faster in its presence, and so on, and so on. Naturally, cancer biologists were very much part of this buzz, given the central role of cell growth in their particular obsession, and what many of them had focused on was how entry into the cell cycle was regulated. Clearly, if an oncogene could push cells into cycle when they were supposed to be resting, this was a property that could cause inappropriate growth, and maybe lead to cancer.

Normally, in the body, the majority of our cells are not in cycle; this is the basis of cancer therapies that attack only actively dividing cells—most cells are untouched because they are quiescent. The resting cells only leave their quiescent state if they detect that the body is wounded or traumatized in some way. They receive these damage messages through receptors on their surfaces, which detect growth factors released from cells in response to wounding. In tissue culture, cells divide nonstop because the growth medium in which they live contains serum, derived from animal blood, which is a very rich source of the growth factors; effectively, tissue culture cells have been tricked into believing that they are permanently wounded and therefore must cycle constantly. To perform experiments mimicking the transition from quiescence to cycling in tissue culture, the trick is to starve the cells of serum, lulling them back into quiescence, a state known as G_0 (G zero). Cells in G_0 can then be studied as they go back into cycle to see which proteins are being made, and in what order. Many experiments in multiple laboratories had by the late 1980s established a hierarchy of genes being switched on upon entry into cycle, with the first out of the blocks being named the Immediate Early genes (followed, unsurprisingly, by the Early, then the Delayed Early genes, with the sad, one plimsoll undone and baggy-shorted Late Genes tagging along at the back). Cancer biologists were all very gratified to find that many of their favourite oncogenes, including *MYC*, were Immediate Early genes, strengthening the idea that oncogenes push cells into cycle. Overexpression of many of these Immediate Early oncogenes was indeed enough to get cycling to occur, and again, *MYC* obligingly fell into line in this respect.

Using Hucky's MYC-expressing stable cell lines, and subsequently, in cell lines they'd made themselves, the Evan lab began to look more closely at MYC's role in the cell cycle. They soon noticed that *MYC* was not a typical Immediate Early gene. Most Immediate Early genes come on rapidly after serum stimulation and then switch off equally rapidly, but *MYC* came on fast and didn't switch off. Instead, although the amount of MYC protein reduced, it lingered on well into subsequent stages of the cell cycle. Was MYC perhaps doing something different to regulate the cycle?

Gerard and Trevor, together with a postdoc, Cathy Waters, decided to set up some experiments looking at what happened if MYC was removed at different points in the cell cycle, not just in G_0. To do this, the protocol was to serum-starve control cells and matching stable cell lines carrying extra MYC for a day or so (over the weekend was a handy time), put the cells back into cycle by adding serum, and then do their assays. They got some interesting results, showing that MYC was indeed needed at a different point in the cell cycle from the rest of the Immediate Early gang, but the experiments were severely hampered by a very awkward technical problem: although the control cells without any extra MYC in them seemed to be fine without serum, the MYC-containing cells were completely unreliable, often dropping dead in such numbers that there weren't enough to analyse at the end of the experiment.

Trevor, Cathy, and Gerard were not particularly fazed by their annoying technical glitch, because the problem that they were dealing with was well known to MYC biologists. Some years earlier, Bob Kingston, a molecular biologist at Harvard, had noticed the same phenomenon, which he had labeled "sick of MYC," and the field had been battling against it ever since. The general assumption was that MYC was such a powerful cell cycle regulator that its overexpression could throw cells into cycle in the wrong way, causing a catastrophically toxic reaction that resulted in death. Indeed, when the Evan lab's errant cells were examined under a microscope, there were always lots of round cells floating in the medium instead of attached to the bottom of the tissue culture flask, looking a lot like cells in the process of mitosis, the final splitting into two new daughter cells. Perhaps a proportion of these cells were unable to complete the cell cycle and died? Furthermore, the tissue culture facilities at Dominion House were primitive in the extreme, prone to temperature fluctuations, contamination, and general inefficiency, and everyone knew that keeping cells alive in such conditions was a tricky task. (Indeed, in the hot summer of 1990, the lab heated up so much that the incubators malfunctioned, killing everybody's cells, and precipitating the closure of the Dominion House laboratories and a wholesale move to Lincoln's Inn Fields, much to everyone's relief.) Experiments struggled on for a month or so, until Gerard suddenly realised that none of this made much sense—if even a proportion of these cells was able to cycle, how come there were always a lot fewer cells at the end of the experiments than there had been at the beginning?

Up to this point, the cells had only ever been looked at under a conventional light microscope, where they could be counted, but where little else in the way of observation was possible. To see exactly how the cells were dying, something far more sophisticated was required. Fortunately for Gerard, it was available at the Lincoln's Inn Fields laboratories: time-lapse cinemicroscopy.

Cinemicroscopy was a very recent arrival at ICRF and, indeed, in the science world, coming into existence at roughly the same time as Gerard's relocation to Dominion House. It revolutionised the way people could monitor cells growing in culture, by allowing them to watch the behaviour of individual cells over time, rather than just look at two time points, the start and the end of an experiment. It was the forerunner of today's incredibly sophisticated time-lapse digitally recording microscopes, but was a very simple setup, consisting of a microscope, a flask full of cells, and an attached cinefilm camera. Every two min or so, the camera was rigged to wind on a frame and take another photo down the microscope. At the end of an experiment, the film was sent off to be developed. Upon its return, it could then be played backwards and forwards endlessly whilst the positions and fates of cells in the field of view were examined and mapped manually.

Gerard made a date with Chris Gilbert, who ran the ICRF facility, and took along some of his cells to be filmed. What he saw changed the course of his scientific life forever: to his astonishment, none of the strange, rounded, floating cells were dividing. Instead, they were all dying, and they were dying spectacularly. Like a clip from a cellular horror movie, the cinefilm showed individual cells detaching from the culture dish and rounding up, then beginning to writhe and contort, their surfaces bubbling as though they had become molten lava, until finally and irrevocably, they exploded into tiny fragments. But what on earth was going on? This death looked very stylised, not like the raggedy, haphazard actions of a confused cell pooping out through mitotic bewilderment. Upon his return to Dominion House, Gerard bumped into Mike Owen, one of the other group leaders there, and got talking to him about the peculiar phenomenon he had just observed. Mike listened, looked at some of the photos, and realised that the cells' behaviour looked strangely familiar. An immunologist, he had a long-term collaboration with an unrelated Owen, John, in Birmingham, who was developing methods for culturing the thymus, the organ from which the white blood cells called T cells originate, in tissue culture. John Owen had documented the appearance during his organ cultures of a large number of strange cells with very dense nuclei, which he had shown were dying by a process first documented almost 150 years previously and named *apoptosis* in 1972. Mike saw that Gerard's cells, although they were fibroblasts, a completely different cell type from those in the organ cultures, were condensing and dying in a strikingly similar way.

The immediate response of a scientist when faced with a new fact to research is to hit the library big time. Therefore, Gerard descended on the dusty stacks of the ICRF library and did some frantic self-education on apoptosis. What follows here is a brief summary of what he found.

The process of apoptosis remained at the level of a mildly interesting phenomenon for many years after it was first noticed in 1842 by Carl Vogt, an inflammatory German scientist and politician more notorious for his subsequent theories on race and his potential role as a Bonapartist spy in Switzerland, than his work describing "natural cell death" in the tadpoles of midwife toads. The first detailed morphological description of apoptosis crops up in 1885, in the beautiful drawings of the eminent anatomist and father of mitosis research, Walther Flemming. Flemming's drawings of "chromatolysis" show ovarian follicular cells shrinking, their nuclei fragmenting, and the formation of apoptotic bodies from their corpses. His discovery of what are still regarded as three of the hallmarks of apoptosis vanished into the ether for many years, together with the work of other 19th century anatomists who identified similar types of cell death, but were keener on studying cell division. In 1951, Alfred Glücksmann revived some interest with a comprehensive review article, in which he convincingly laid out the case for cell death being required for the sculpting of the body during neonatal development, the development of organs, and for metamorphosis in amphibia and insects. Richard Lockshin and John Saunders took this work further in the 1960s, when they realised that developmental cell death had to be tightly controlled, with a biochemical mechanism behind it, and a corresponding set of genes involved. Lockshin and his PhD supervisor Carroll Williams coined the phrase "programmed cell death" to reflect their view of the regulated nature of the process. Again, although embryologists now took on board the concept of death being a regulated part of development, this work did not penetrate the mainstream scientific consciousness.

The difficulties that death had had in establishing itself as a proper biological event worthy of study began to be surmounted after 1972, following the publication of work by an Australian pathologist John Kerr, and his collaborators in Aberdeen, Andrew Wyllie and Alastair Currie. Kerr had rediscovered Flemming's hallmarks of apoptosis in injured liver tissue, calling it "shrinkage necrosis," and whilst on sabbatical in Aberdeen, he, Wyllie, and Currie published a paper characterising, in multiple tissue types and cells both during normal development and in disease, the stages and morphology of this form of death. James Cormack, Professor of Greek at Aberdeen, suggested the term *apoptosis*, meaning the dropping of petals from flowers, or leaves from trees, for Kerr, Wyllie, and Currie's paper. As with Lockshin's earlier terminology of programmed cell death, the name was chosen to imply that the process was part of a cycle of life and death and was regulated, rather than haphazard. (Unfortunately for Professor Cormack, his proper Greek pronunciation, "Appo-TOE-sis," has morphed in the mouths of most of the scientific community into "Appop-TOE-sis," with only a few die-hards remaining true to the original.)

Although Kerr, Wyllie, and Currie had come up with a cracking name and also speculated on the importance of apoptosis in multiple processes, including cancer, their work attracted little interest, partly because there seemed no good way of nailing down exactly what mechanisms cells used to initiate and execute apoptosis; it still remained very much a process to be described, rather than understood. However, the advent of two key papers in two disparate fields, immunology and worm genetics, finally provided biochemical and genetic ways into the problem. For the biochemists, Andrew Wyllie published a paper in

1980 looking at the induction of apoptosis in thymocytes, the same cells that Mike Owen's collaborator John Owen was studying. Wyllie showed that an enzyme called an "endonuclease," which chops up DNA into small pieces, was needed for apoptosis to occur, and best of all, showed that one could monitor the activity of the endonuclease very easily, by looking on a size-fractionating gel for a ladder of DNA, where each of the rungs of the ladder were separated by exactly 146 bp. (The endonuclease cuts at gaps between nucleosomes, the structures onto which chromosomal DNA is wound, and each nucleosome accommodates 146 bp of DNA.) Nucleosome laddering gave scientists an assay for early apoptosis that was both easy and quick, rather than the laborious and unquantifiable pictures produced by the histopathologists and embryologists from cut-and-stained sections of tissues.

On the genetics side of things, Bob Horvitz and John Sulston, working in Cambridge on tracking the developmental origins of every cell in the adult nematode worm *Caenorhabditis elegans*, realised that of the 1090 cells generated during development of an adult hermaphrodite worm, the same 131 cells always died, and moreover, they died by apoptosis. *C. elegans* is a geneticist's dream because it grows rapidly and can be mutated easily, and Horvitz and Sulston's discovery meant that it was now possible to start looking for mutant worms in which cells meant to die stayed alive, or vice versa. The underlying genetic mutations causing "undead" phenotypes would be in genes controlling and executing apoptosis, and these genes could be isolated, cloned, and their DNA sequences determined. The importance of this work was recognised when Horvitz, Sulston, and Sydney Brenner (whose foresight had driven the development of *C. elegans* as a model organism) won the 2002 Nobel Prize for their discoveries concerning the genetic regulation of organ development and programmed cell death.

By the time Gerard had his momentous conversation with Mike Owen, a total of around 500 papers had been published about apoptosis and programmed cell death, mostly in the fields of developmental biology and immunology, with a few also looking at the behaviour of cancer cells. Many were still descriptive, but the most recent crop of papers was starting to identify genes involved in the death process and to try to put these in the context of overgrowth, as occurred in cancer. One of these genes, *BCL2*, caused Gerard's scientific radar to start beeping madly because it was already very familiar to him, but in a different context: *BCL2* had been suggested to be an oncogene.

The *BCL2* gene first cropped up as a candidate oncogene in 1984. Since then, it had been very resistant to attempts to prove it really caused cancer because it didn't seem to work in any of the standard assays, such as causing cell overgrowth. However, in 1988, David Vaux, a postdoc in Suzanne Cory and Jerry Adams's lab in Melbourne, convincingly showed that the BCL2 protein appeared to be a different sort of oncogenic animal, namely, an anti-apoptotic survival factor. It couldn't force cells to cycle and make tumours that way (in fact, it arrested the cell cycle), but instead it allowed distressed cells to stay alive until they could be taken advantage of by a second, growth-promoting oncogenic mutation. In the Vaux paper, the second cooperating oncoprotein able to work with BCL2 was none other than … MYC.

The story at this point might just have petered out into another publication in a worthy, second-rank journal, with some dutiful observations, albeit using a cute new technique, about the mechanism by which MYC caused the cell death that so many researchers in the field had already encountered. Indeed, the atmosphere in the burgeoning field of molecular biology at that time was exactly conducive to such an outcome. Almost all of those working on deciphering the ways in which proteins and genes functioned were basic researchers, doing bottom-up, technically challenging science, in many cases still at the dawn of knowledge in their particular fields, where feverish excitement could be generated merely by finding out how protein A might connect to protein B, and how A and B interacted with C. All was unknown, even at the lowest, nuts-and-bolts level. Furthermore, there was a deep and snobbish divide between such research and what in today's parlance would be called translational research: Taking basic science observations and considering whether they would be useful in a clinical context was widely considered to be the easy option, for those who could not generate new ideas or data but spent their careers recycling the observations of others.

These factors conspired to produce a culture in many labs of single-minded dedication to the problem at hand, with little thought for any potentially wider implications. Gerard, however, had an edge on many of his colleagues and competitors: He had trained in two places with world-famous reputations for omnivorous scientific reading and thinking. At the Laboratory of Molecular Biology in Cambridge, he had done a PhD in a clinically oriented lab, trying to find out whether the body's immune system was able to detect and kill cancers. At the same time, he had been exposed to the sometimes frighteningly adversarial but intellectually invigorating atmosphere there, where the opinions of the lowliest PhD students were considered as valid as those of the most eminent Nobel laureates, as long as they were logical and thought-provoking. Secondly, Gerard's postdoc supervisor, Mike Bishop, was and is one of the most perceptive thinkers in cancer biology, and, having originally trained as a medic, naturally viewed the study of the molecular biology of cancer as a means to an end—understanding cancer itself. An even more omnivorous reader than Gerard, Mike also introduced him to the writings of Stephen Jay Gould, and hence to evolutionary theory, converting Gerard to his view that all biological processes must be evolutionarily explicable and that, therefore, any theory that does not make evolutionary sense is unlikely to be the correct one.

With his science brain configured in this slightly unusual way, Gerard made a conceptual leap that to him seemed blindingly obvious, linking his discovery that MYC could cause apoptosis to that of David Vaux showing that MYC could cooperate with BCL2. Death caused by MYC was a programmed part of the cell's lifestyle, not a consequence of a mitotic car crash; an executed decision, not an accident. This meant that MYC was constituted to have dual, seemingly opposite functions, able to push cells into cycle, but also forcing them to die under certain circumstances. BCL2, on the other hand, actively stopped cells from cycling and protected them from death. Might not this mean that the whole idea of oncogene cooperation, originally proposed by Parada, Land, and Weinberg to be a simple alliance between one oncogene driving one aspect of tumour formation, such as too much cell cycling, and another oncogene driving a different aspect, such as stopping death, was far too simplistic? Better, surely, was a theory in which cooperating oncogenes not only had complementary cancer-causing functions but were able to override the other's inbuilt safety mechanisms, delivering a far more potent double whammy to the normal cell? Normally, if a cell accidentally made too much BCL2 it wouldn't matter because it wouldn't be able to cycle, and if it made too much MYC, that wouldn't matter either because it would just die. Only in a situation in which the two oncogenes were wrongly switched on together would MYC override BCL2's cell cycle block, pushing cells to cycle, and BCL2 stop the apoptotic signals being sent by MYC from being executed, and only then would there be a possibility for a cancer to initiate.

Gerard's new idea had enormous implications for cancer biology as a whole. Since the discovery of the potency of individual oncogenes and their abilities to make cells transform into immortal, continually growing lines in culture, there had been puzzlement over why, if oncogenes were so powerful, people did not develop tumours all the time. After all, humans have around 100,000 billion cells in their bodies, so the chances of a cancer-causing mutation occurring are huge. Why should normal cellular proliferation be so easy but cancerous growth so hard? According to Gerard's thinking, the answer was that pathways that drove growth were configured so that if one got activated out of context, there was a tumour-suppressor defence mechanism evolved into it. The decision for a cell to proliferate could be compared with releasing an elaborate combination lock: Under normal circumstances, the code for the lock would be figuratively provided by a set of growth signals activating a particular set of receptors on the cell's surface. If the combination was correct, the lock would open, unlocking the proliferative capacity of the cell and putting it into cycle. If the signals were of the wrong sort, or in the wrong order, or were too strong, the code would be wrong and the lock would refuse to open. The radical difference between the lock analogy and what happens in the body is that rather than the lock just staying shut, any tampering with the mechanism results in it blowing up, destroying the cell in the process.

The experiments for the paper Gerard and his coworkers eventually published in *Cell* on 14 January 1992 are simple and to the point. Evan et al.'s "Induction of Apoptosis in Fibroblasts by c-Myc Protein" contains eight figures, illustrating the key results. All of the data are from experiments performed in immortalised Rat-1 fibroblasts, a type of cell derived from connective tissue, and one of the workhorses of tissue culture. The first two figures in the paper show the original observations that the lab made, that when fibroblasts overexpressing MYC protein are starved of serum, they are forced to cycle repeatedly, in contrast to the control samples, which drop out of cycle. Despite the continual cycling, the MYC-expressing cells do not increase in number. Figure 3 is the core of the paper, showing that in the presence of MYC, the cells are dying by apoptosis. This is shown using stills from the cinemicroscopy films, a photo of a gel with the apoptotic nucleosome laddering pattern, and some very pretty electron micrographs showing individual cells at different stages of apoptosis. Figure 4 uses cells containing an artificially modified MYC protein that can be switched on and off by addition of the hormone estradiol to the cultures, and is there to show that apoptosis only happens when MYC is switched on, showing that MYC protein really is what is driving the apoptosis. Figure 5 answers the question of whether all of the data are simply a horrible tissue culture artefact only observable in one particular cell line, by showing that MYC is also able to induce apoptosis in serum-starved fibroblast cells taken directly out of an animal. Figure 6 looks at a set of cell lines with different amounts of MYC in them and shows that the number of cells dying in culture depends on how much MYC is around, with a complete correlation between more death and more Myc. Finally, Figures 7 and 8 show that MYC can induce apoptosis in any arrested fibroblast, irrespective of how the arrest was achieved, and, using the switchable MYC mutant inducible by estradiol, whether or not MYC was present when the arrest started. Finally, there is a table showing that the same parts of the MYC protein known to be important for its activity as an oncogene are also required for its ability to induce death.

The simplicity of the experiments, and the carefully discussed conclusions, belie the frantic activity that had ensued following the first observations the lab had made. Almost the first step that Gerard took was to make an appointment to go and visit Andrew Wyllie, one of the fathers of apoptosis, who had moved from Aberdeen to be Professor of Experimental Pathology at Edinburgh University. Gerard's initial conviction that he had stumbled onto something important was confirmed by Wyllie, who could see the enormous implications of apoptosis being a mainstream suicide mechanism rather than an occasional habit of white blood cells and developing embryos. Wyllie became a valued colleague, expert in the arcane lore of the apoptosis field, and was responsible for suggesting to Gerard that the work was so interesting that it should be submitted to *Cell*, for maximum visibility. Back in the lab, Trevor, Dave, and Cathy were working all hours doing endless tissue culture, sometimes staying all night. Everyone was excited because they all realised that, finally, after all of the awful things that had happened to the lab in the last few years, they had got their big break. If MYC causing apoptosis was as big a deal as they thought, they were set for stardom, or at least temporarily released from the fear of failure that haunts the hearts of all scientists.

Others in Gerard's circle of friends and collaborators had also realised how much this mattered, both to science and to Gerard's prospects. Experiments were helped along by the continued collaboration with Hucky Land's lab, and an awful lot of talking with anyone and everyone willing and interested enough to listen. The science grapevine was quickly alerted to the new work as people at conferences spread the word, and Gerard started to receive a trickle of invitations to give talks at meetings that quickly transformed into a flood.

To begin with, many cancer biologists reacted with extreme scepticism to the idea that an oncogene, especially one so powerful as *MYC*, could be hard-wired to cause a cell to kill itself. Gerard recalls the first talk he ever gave about it, at a Gordon conference, traditionally a venue for the frank exchange of views and a very prestigious gig for an up-and-coming scientist. Leaving aside the several people who walked out, whether from disbelief or preferring the competing charms of an early coffee break, many prominent researchers in the field were reluctant to let go of the idea that this property of MYC's was no more than a by-product of its forcing cells into cycle. The consensus view was that an individual cell was able to sense its health or otherwise, and if it detected that something bad was happening, such as excessive cycling, it would fall sick and die, by a theoretical mechanism snappily named "mitotic catastrophe." Therefore, the phenomenon "Sick of MYC" was just that: a terminal illness, a set of conflicting signals whereby Myc said one thing to the cell, the rest of the cell thought something else, and the two viewpoints were irreconcilable.

The problem with such anthropomorphic thinking was that, viewed from an evolutionary perspective, it made no sense. Firstly, the notion that an oncogene could have a main function, from which all its properties flowed, was nomenclature-based determinism, a belief that the first thing a gene was discovered to do had to be the most important. Just because MYC was first noticed in the context of excess cycling did not mean that that was the main thing it did. More importantly, evolutionary theory now held that the organism was the fundamental unit of evolution, with its genes just a vast collection of biological meccano that could be assembled into as many different networks as were needed for the body to function and survive; genes had a function, but that function could be used in a whole organism for multiple purposes. Viewed like that, it was perfectly reasonable for MYC to be involved in any number of processes and that these processes might dictate conflicting outcomes was irrelevant: The downstream readout was a function of the pathway as a whole, not its individual components.

The second big obstacle to acceptance of Gerard's work was that the idea that normal cells were programmed to commit suicide if tampered with, that they were humble foot soldiers killed off to protect the greater cause, was very much a novelty at the time. Cells were viewed as important nonexpendable entities, wired to stay alive and reproduce themselves at all costs; there was an almost emotional attachment in the field to the life of an individual cell. However, such sentimentality was not shared by the bodies of those subscribing to this view; while they grew in the womb, their cells had been killed with ruthless abandon, with apoptosis used as a developmental tool to sculpt the body, editing out, amongst other things, the embryonic webbing between their fingers and toes. What was true in development was also true in immunology, where vast swathes of immature cells died on their way to maturity, because to do otherwise opened up the possibility of autoimmune disease, when rogue cells start to destroy the body's own tissues. Interestingly, MYC had already been implicated in apoptosis of immune cells, in a 1991 publication from John Cleveland's lab at St. Jude Children's Research Hospital in Memphis, Tennessee, and a second paper showing that loss of MYC could protect mature activating T cells from apoptosis appeared in 1992, from Doug Green's lab in La Jolla, California. These papers were also greeted with some scepticism, but the general view was that because immune cells were "supposed" to kill themselves, it was less off the wall than Gerard's work in fibroblasts, where apoptosis was almost completely undescribed.

Other objections continued to do the rounds, one of the most persistent being that if apoptosis were so all-pervasive, why wasn't it more obvious? As Gerard put it in a 1974 article, "Even a cursory inspection of your nearest colleagues will indicate that they comprise palpably living cells with … little evidence of substantial cell death" (Evan 1994). This point was cleared up once a harder look was taken—it turns out that the speed with which apoptosis happens and with which the debris is cleared up by other cells is simply so fast that it had been missed—but it hid an understandable and very human grievance on the same subject: All of Gerard's really rather good colleagues in the MYC field were having to admit to themselves that in their obsession with MYC and the cell cycle, they had all failed to see the apoptotic elephant in the room, which was something of an assault to their collective scientific pride. Fortunately, all was (mostly) forgiven very quickly, once it became clear that Gerard really was on to something; the apoptosis bandwagon was rolling, and everyone wanted to hop on board.

Science, like pop music, has its fair share of one-hit wonders; Gerard was well aware that if he didn't capitalise on his discovery, he wouldn't get another chance at success. Once his first observations were published, he knew that he would have to operate at a flat gallop if he wasn't to be flattened in the stampede. He was simultaneously terrified that he'd mess up and elated that he'd finally found his métier. Proving that apoptosis had a central role in cancer, finding out how that role might be fulfilled, and using the knowledge gained to design anti-cancer therapies was about as intellectually satisfying a task as he could conceive of. He and his lab leapt into the fray, knowing that what was great about his big idea—that apoptosis was a programmed safety mechanism carried at the heart of all cells—was that it was eminently testable and experimentally very tractable. Matters started to move very quickly. Over the next few years, the Evan lab became a paper mill; as the lab filled with ambitious and hard-working postdocs and graduate students, lured by the promise of being in on the start of something, every experiment that they did seemed to work, and all of their papers got into highly respected, high-profile journals.

In the autumn of 1992, Gerard published a paper in *Nature* about oncogenic cooperation between MYC and BCL2. As he'd surmised, the two oncogenes were able to work together because they could compensate for each other's weaknesses: BCL2 protected cells from MYC's murderous tendencies, whereas MYC chivvied cells into cycle, overcoming BCL2's cell cycle block. However, the paper was more than just a vindication of Gerard's original thoughts; the Evan lab had been caught up, and their publication was back-to-back with another saying the same, from the lab that had shown the link between MYC and apoptosis in T cells—that of Doug Green in La Jolla. Events could have taken a very nasty turn, given the highly competitive nature of the subject, but fortunately for both labs, Doug and Gerard soon met, at a conference in the windswept wastes of the University of East Anglia campus. Doug, heading back into the conference hall following a quick trip to the lavatory, met Gerard heading out on the same mission (an awful lot of coffee is consumed at meetings). The two men realised pretty quickly that despite being competitors, they were so similar in temperament and outlook that they couldn't help but be friends first. Happily, they've been best friends ever since: mutual repositories of scurrilous stories, enthusiastic occupants of numerous conference bars, and scientifically distinguished collaborators and competitors.

The evidence for a central role for apoptosis continued to pile up rapidly. Other oncogenes and transcription factors, notably E2F and E1A, were shown by other labs to have the same dual nature as MYC, making commonplace what had once seemed revolutionary. Earl Ruley's lab showed that RAS, another very unpleasant oncogene mutated in one in three of all human cancers, could induce replicative senescence, when cells exit the cell cycle but do not die, showing yet another control mechanism cells use to fight oncogenic changes.

In 1994, in a paper in the *EMBO Journal*, the Evan lab proved the most important tenet of Gerard's theory, that of MYC's duality. In normal conditions, a cell exists within a network of social survival signals, which suppress MYC's signals to the apoptotic programme but allow its proliferation signals to make it through, so that cells grow. If the survival signals are removed, MYC's apoptotic signal is no longer suppressed and cells die. However, if the survival signals are maintained artificially, MYC cannot kill the cells but drives them into cycle, causing tumours. By the time this paper came out, however, the battle for acceptance was long won, and the field had settled down to the long slog of working out the exact mechanisms involved in triggering and then executing apoptosis.

Today, we know an awful lot about apoptosis—mechanistically, it is probably one of the best understood cellular processes, thanks to the combination of biochemistry and genetics that proved so important in the genesis of the field and have continued to have a vital role. From an average of about 40 papers a year pre-1992, there are now close to 10,000 articles published about apoptosis annually, more than one every hour. However, despite roughly one in 50 of these publications having some mention of MYC in them, we are still not quite sure exactly how MYC works to cause apoptosis. What is known then?

MYC is a transcription factor, implicated in switching on some 15% of all genes. It is ubiquitous, and as befits such a powerful protein, highly regulated, coming on only in response to growth signals. Just about every process in the cell appears to depend to some extent on MYC, and therefore, its power does not derive "from any specific subset of its functions, but through its unique capacity to coordinate and integrate the diverse gamut of processes that, operating together, underpin somatic cell expansion" (Soucek and Evan 2010). In other words, MYC is the cell's puppet master. MYC is deregulated and overexpressed in nearly all tumours, and analysis of the genes it switches on (the "MYC signature") gives some insight into its fearsome reputation as an oncogene; the MYC signature looks an awful lot like the transcriptional signature of a stem cell, making it very likely that MYC has a big role in the generation of cancer stem cells. These cells, which carry the cancer blueprint, are often able to survive anti-tumour therapies and are responsible for recurrence of disease.

MYC's ability to cause apoptosis is partially understood, because it can induce the p53 tumour-suppressor protein, causing cells to kill themselves. However, there are clearly MYC-driven apoptotic pathways independent of p53, about which we currently know very little. Rather than being a sole regulator of apoptosis, as it was in the artificially created tissue culture milieu of Gerard's original experiments, in normal life MYC functions to sensitise cells to a huge range of pro-apoptotic insults, including starvation, oxygen deprivation, DNA damage, and the activation of death receptors—proteins able to transmit apoptotic signals coming from outside the cell. In this duality, it shares a common function with many proto-oncogenes, acting as a driver of proliferation, but also as a sentinel detecting unnatural changes to a cell, nearly always brought about by tumour-specific growth.

Anti-cancer therapies aimed at reawakening a cancer cell's lost ability to apoptose have been pursued with much energy over the past decade. A 2002 review, written for the inaugural issue of the journal *Cancer Cell* by Gerard and Doug Green (they concocted it over a shared Thanksgiving weekend, although thankfully, no traces of turkey-based thinking seeped into the final product), pitched the idea that cancers are built on a precarious platform of two mutations, one causing overgrowth and the other causing apoptosis suppression, and that if one or the other of these mutations were to be counteracted, the cancer would die. The hypothesis flows directly from the original insights gained from the interaction of MYC and BCL2, and to general excitement, it has, indeed, proved possible to topple a MYC-driven cancer by knocking out its founding MYC mutation. Gerard's lab, the lab of his old mentor Mike Bishop, and that of another ex-Bishop postdoc, Dean Felsher, have engineered mice with a switchable *MYC* gene embedded in their genomes. When the mice are treated with the switch compound, MYC comes on, causing tumours, but these tumours regress rapidly as soon as MYC is switched off again.

These experiments challenge the idea that the founding mutations of a cancer are almost irrelevant once the tumour has got going and suggest that finding out how exactly a tumour has come by its resistance to apoptosis, and correcting the causative mutation, could be the Achilles heel of cancer. Could cancer be a disease that has such a florid outcome that its fundamental simplicity is masked? This is not so outrageous because there is a strong analogy to bacterial infection; the multiplicity of vastly different illnesses caused by bacteria can all be cured by one drug class, the antibiotics, which attack the fundamental weakness of all bugs, their dependence on an intact cell wall. As with antibiotics, combinatorial therapy will be required to cure cancers, to ensure that resistant substrains are also killed, but the prospect of anti-apoptotic therapy for cancer is a very hopeful one at present.

And what of Gerard? After all the years of striving to be successful and shaking off the traces of his Cambridge experience at the Ludwig Institute, he realised that, actually, it wasn't that much fun being a world authority on something, because many people stopped arguing with him, assuming that he knew what he was on about. It was also no fun being part of the furniture of an institution, and so, after a very productive decade at the ICRF, Gerard left to go back to San Francisco, to the new UCSF Cancer Center, where he started again at the bottom of the heap, working on mouse models of cancer and driving a rather snazzy little red sports car across the Golden Gate Bridge to the lab every day. Having worked his way up to become a major player in both the MYC and p53 mouse worlds, he moved back across the Atlantic in 2009, to become Head of the Cambridge Department of Biochemistry and Sir William Dunn Professor. Gerard's current research is focused on the processes responsible for the genesis and maintenance of cancers, in particular, cancers of the pancreas, colon, brain, skin, and liver. Understanding the molecular mechanisms that underlie the cell suicide machinery and how it can be manipulated therapeutically is still the overarching aim of his laboratory, and he continues to be one of the most productive and thought-provoking cancer biologists working today.

For his work on apoptosis and cancer, Gerard was elected to the UK Academy of Medical Sciences in 1999 and became a fellow of the Royal Society in 2004. His 1992 *Cell* paper is an acknowledged classic, having been cited more than 2600 times, and remains the most highly cited research paper ever to have been published by scientists working at the London Research Institute.

Web Resources

www.youtube.com/watch?v=G0D29CO2Lq4 Gerard talks about his research.

www.youtube.com/watch?v=xdLPpdoU2Nc Video of cells undergoing apoptosis.

Further Reading

Bishop JM. 2003. *How to win the Nobel Prize.* Harvard University Press.

A wonderful autobiography by Gerard's postdoctoral mentor Mike Bishop, the 1989 Nobel laureate in Physiology or Medicine.

Green DR. 2011. *Means to an end: Apoptosis and other cell death mechanisms.* Cold Spring Harbor Laboratory Press, Cold Spring Harbor, NY.

Technical but entertaining book by Gerard's drinking partner Doug Green. There's a companion website too: http://celldeathbook.wordpress.com.

Quotation Source

Evan GI. 1994. Old cells never die, they just apoptose. *Trends Cell Biol* 4: 191–192.

Soucek L, Evan GI. 2010. The ups and downs of myc biology. *Curr Opin Genet Dev* **20**: 91–107.

CHAPTER 8

Walk This Way

Sex determination, how an embryo decides to become male or female, is such a fundamental process that it comes as a surprise that we had no clue how it worked until a little over half a century ago; in 1959, in a great example of scientific serendipity, the beginnings of the answer were born out of the boredom of a 24-year-old researcher in Edinburgh, Pat Jacobs.

Pat's expertise was in cytogenetics, the analysis of chromosomes, and in the wake of the 1950s nuclear arms race, she was hired by a small MRC-funded group to compare the chromosomes from patients with radiation-induced leukaemias with those with normal leukaemia. It was a laudable project, but unfortunately, Pat couldn't find a single example of a radiation-induced leukaemia and had to mark time making a larger and larger control sample of normal leukaemic chromosome spreads. Therefore, when her boss Michael Court Brown introduced her to John Strong, an endocrinologist who wanted some chromosome spreads performed from one of his patients, she jumped at the chance to break the monotony. Strong's patient had been classified as suffering from Klinefelter's syndrome, a sex-reversal disorder then held to result from a female with two X chromosomes somehow becoming a male. However, Pat noticed something rather odd when she looked at the chromosome spreads; there were definitely two X chromosomes, but instead of the normal human count of 46 chromosomes, there were 47. Even more weirdly, the interloper was small, and to Pat's practised eye, looked an awful lot like a Y chromosome. Because the Y chromosome was popularly held to be a piece of functionless junk, this seemed odd, but analysis of other Klinefelter's patients yielded the same result—all had two X chromosomes and an additional Y.

Her youth and relative naivety meant that it wasn't until Pat showed the data to Michael Court Brown that she had any idea of its huge significance, but Court Brown realised immediately that what he was looking at was going to overturn all notions regarding how human sex was determined. Jacobs and Strong's paper, published in *Nature* in 1959, was the first demonstration of the binary nature of sex determination in mammals—if there is a Y chromosome, you make boys, and if there isn't, you make girls. However, it would take until 1990 until it was finally known what it was on the Y chromosome that was responsible for making the decision. It would fall to another researcher who had wandered into the field whilst marking time waiting for reagents, to make the discovery. This time, however, the path to success had many more twists and turns than ever encountered by Jacobs and Strong, and the competition proved to be unhealthily intense.

In the 30-odd years between the Y being fingered as the source of the Testis Determining Factor, or TDF (developed testes being the physiological criteria for maleness), and the discovery of what exactly on the Y was responsible, science had to cover a lot of ground. In 1959, genetics, the study of how characteristics are inherited, had accumulated a venerable history, but the gap between the statistically derived maps of linked genetic loci, based on the frequency with which two loci were inherited together or separately, and the physical reality of the chromosomes, where the genes corresponding to the loci were actually located, was enormous. How did one superimpose the abstract linkage maps onto the DNA? Those working in the field of human genetics had an additional problem, because they were unable to set up the breeding experiments that were the lifeblood of genetic research in lower organisms, and could only make very basic linkage maps based on a few family pedigrees.

Fortunately for the human geneticists, in 1960, a group in France made a discovery described by J.B.S. Haldane as a substitute for sex, possibly the first and only time that such an idea has been viewed with unbridled enthusiasm. The idea of mammalian somatic cell or parasexual genetics, as it was more prosaically called, was not new, having been proposed by Guido Pontecorvo in the 1950s. Pontecorvo was one of the first microbial geneticists, hunting for fungal genes by mutation and recombination, and had noticed a rare spontaneous event in his favourite fungus *Aspergillus*

Guido Pontecorvo. *(Photograph courtesy of CRUK London Research Institute Archives.)*

nidulans, in which nonsex (somatic) cells fused together and generated offspring bearing characteristics of both parent cells, circumventing the need for mating. Ponte, as he was universally known, realised that if somatic animal cells could be made to fuse in this way, vast genetic opportunities would open up; unfortunately, Ponte does not occupy the place in history he deserves for this fundamental insight, because he did not succeed in getting mammalian cell fusion to work. However, this had now been remedied. Ponte must have been very unlucky, because the paper published by Serge Sorieul, Francine Cornefert, and Georges Barski was apparently very straightforward.

Sorieul, Cornefert, and Barski's method for somatic cell fusion was based on tissue culture. They reported that if they mixed two different sorts of cells together, the resulting mixed cultures contained cells with characteristics of both parental types. With Boris Ephrussi, Sorieul showed in 1962 that the cells had in effect "mated"—their chromosomes had mixed together and then resegregated to create offspring carrying a genetic contribution from each parent. Most excitingly for geneticists, in 1967, Mary Weiss and Howard Green found that if mouse cells were mixed with human cells, the battle of which chromosomes were retained by the hybrid interspecies offspring was almost invariably won by the mouse chromosomes.

With the advent of somatic cell hybrids, as they were called, it was possible to derive cell lines whose chromosome complement included just one human chromosome, with all the rest being mouse. These could then be used to map loci to a particular chromosome, as long as one had an assay for the locus of interest. Panels of cell lines carrying different human chromosomes could be screened for the presence or absence of a human gene product, and hence its parental gene. Experiments were relatively simple: Set up your assay, whatever it was, on the 20 or so cell lines, each with their single human chromosome, and see which line comes up positive. Check which chromosome is in that cell line, and bingo! You've mapped another gene. Combined with the fact that breeding was no longer necessary, experiments took days rather than years, and in the decade following, somatic cell hybrids were used to map a few hundred genes with known and assayable gene products onto specific human chromosomes, including the X.

Sadly, for those interested in the Y chromosome, the pickings were still very thin, even using somatic cell genetics. The problem lay in the fact that, unlike the X, which was positively bursting with genes, the Y chromosome seemed to be an echoing wasteland—none of the known genes seemed to map there at all. This emptiness, however, fitted very neatly with the prevailing hypothesis regarding sex determination, which held that the default fate of an embryo was to become female, and this could be diverted into maleness by the expression of just one gene. Such a big decision, it was argued, should be a simple on/off switch, a binary decision, to leave as little room as possible for mistakes. An interesting, if completely wrong, idea arose suggesting that maybe the Y only held one gene, the testis-determining TDF, and that therefore, the only criterion for being TDF was to be Y-linked. A large number of adherents of this idea, who should probably have known better, hared off after the H-Y antigen, the only protein known to be associated with the presence of a Y chromosome, and it was acclaimed as TDF for some 10 years, until the theory was comprehensively demolished in 1984 by the British embryologist Anne McLaren and her colleagues, who showed that the presence or absence in mice of H-Y antigen didn't make a blind bit of difference to whether mice turned out male or female.

After this and other equally misguided attempts to find a gene, any gene, on the Y, the more enlightened amongst the field realised that to identify the genes on the Y chromosome and to find which of them was TDF, was clearly going to need a different technique, one that was independent of what the genes encoded; in other words, the genes would have to be mapped simply by their presence on the Y, not what they did. And, at the end of the 1970s, a suitable mapping technique appeared, provided, inevitably, by the fertile minds of molecular biology.

The molecular biologists' annoyingly practical habit of doing things from the bottom up, starting from the DNA, proved invaluable to the geneticists, struggling with their scientifically indigestible chromosomes. As the 1970s progressed, molecular biology, as related in Chapter 2, began to provide the tools with which to manipulate and analyse both small and large pieces of DNA. Restriction enzymes, able to cut DNA at a particular sequence, made it possible to reduce chromosomes to more manageable, smaller chunks. The fragments could be separated according to size by gel electrophoresis and then cloned and amplified by insertion into plasmid vectors. Chunks of DNA could be searched for genes, using Southern blots, and the sequence of bases of which the fragments were composed could be determined, after the development of DNA sequencing. In theory at least, it was now possible to go from a chromosome to a gene.

As geneticists and molecular biologists alike began to wield their new tools in earnest, they realised that the combination of DNA cloning and the availability of increasingly more sophisticated vectors for carrying larger and larger fragments of DNA meant that it was possible to fragment chromosomes and, indeed, genomes and clone all of the fragments into vectors: DNA libraries could be made carrying complete genomic sequences in manageable chunks. These libraries could be probed with radiolabelled DNA or RNA molecules in the same way as Southern blots, giving another way to isolate genes, but their existence also opened up an even more interesting possibility: If there was a way of ordering overlapping library clones in the same way as a library of books, the result would be a DNA analogue of the genetic maps made by linkage analysis. Finally, the prospect of mapping genes on the Y chromosome simply by their location seemed more than just a daydream, as long as someone could figure out how to order the DNA library clones into a coherent linear array.

The key to the puzzle of how to order DNA libraries was provided in 1980 by the newly appointed Director of the ICRF, the 43-year-old Walter Bodmer. Human genetics by this time was a busy anthill of scurrying activity, and although it seems somewhat implausible to liken him to any type of ant, Walter Bodmer was a central figure in the antheap. His scientific lineage was impeccable. At Cambridge, he had been one of the last students of Sir Ronald Fisher, a genius who was the father of both modern statistical analysis and population genetics, and he had then moved to Stanford to work with Josh Lederberg, one of the founders of molecular biology. He got on so well with Lederberg that he became a faculty member at Stanford, where he made his codiscovery of the Human Leucocyte Antigen (HLA) system, crucial in the body's rejection or otherwise of organ transplants. Bodmer's HLA work led him from his first interests, plant and bacterial genetics, into human genetics, and he returned to Oxford as Professor of Genetics in 1970, where he continued his work on HLA, becoming a world leader in the field that he had founded. Around the time of his arrival at ICRF, he and Ellen Solomon, one of the new faculty he brought

Walter Bodmer. *(Photograph courtesy of CRUK London Research Institute Archives.)*

with him, proposed an idea, suggested at the same time by David Botstein, Ray White, Mark Scolnick, and Ron Davis in the United States, that solved the problem of how to order DNA libraries. (As a footnote, Botstein and his colleagues received rather more recognition than Bodmer and Solomon for their idea, perhaps because the Botstein paper was entitled, "Construction of a Genetic Linkage Map in Humans Using Restriction Length Polymorphisms," and Bodmer and Solomon's was called "Evolution of Sickle Variant Gene." In his defence, Walter always said he didn't bother promoting his idea much because he thought is was blindingly obvious.)

The method proposed in the two papers centred on the discovery of restriction fragment length polymorphisms, or RFLPs ("rifflips"). Sequences in even the most conserved regions of the genome tend to vary slightly between different people, and the slight variations are heritable, meaning that these polymorphisms, often a change in just one nucleotide, can be used as genetic markers. Detecting RFLPs might seem a complex challenge, but fortunately, as the acronym RFLP suggests, biologists had a way of "seeing" small snippets of the genome, namely, the recognition sequence of a restriction enzyme. For example, if in a large fragment of DNA such as those cloned in the genomic libraries, the sequence GCGGCCGC occurred twice, a restriction enzyme called NotI would be able to cleave the fragment twice because GCGGCCGC is its recognition sequence. If the cleaved fragments were then run on a Southern blot and probed with the appropriate radiolabelled DNA, three bands would be seen. (Why three? Cut a piece of string in two places and it becomes evident.) However, if the same fragment was extracted from a different person's genome, it might be that one of the NotI sites was no longer present, owing to a mutation having caused a change, say, of one of the C bases to an A. Therefore, cleavage by NotI would result in only two bands on a Southern blot, and thus one would have a marker, an RFLP, to differentiate between the first and the second person's DNA in that area. So now, because such markers exist in many places in the genome, not just for NotI but for many other restriction enzymes, there was a way of telling the difference between the genomes of two different people. Using RFLPs, the isolation of a locus based on its position, rather than its function, was now possible, and the process acquired a new name, positional cloning, or reverse genetics. RFLPs had huge ramifications for the cloning of disease genes because normal DNA could be compared with DNA from disease carriers, and the faulty region identified. The first locus to be successfully linked to a disease was the region of the X chromosome carrying the Duchenne Muscular Dystrophy gene, in 1982, and this was rapidly followed by many others.

Following on from the idea of using RFLPs as a way of distinguishing between different genomes, the one extra conceptual step taken by Bodmer and Botstein and their colleagues was to realise that RFLP technology could also be used within a single genome to create an ordered map. By comparing the restriction enzyme patterns made by digesting all of the clones in a DNA library made from just one genome, shared patterns, indicating overlaps between different clones, could be detected. Rather like piecing together a panoramic photograph from individual overlapping shots, clones could then be put in order. This simple but clever idea was the seed from which the Human Genome Project grew, culminating in the publication in 2001 of a working draft of the complete DNA sequence of the human genome.

But what of the Y? The advent of positional cloning techniques fired the starting gun in the real race to find TDF, and the runners and riders were an interesting bunch. Two in particular seemed to be worth betting on: in London at the ICRF, Peter Goodfellow, a former Bodmer graduate student recently returned from Stanford, had moved onto the Fifth Floor and was looking for a project with which to make his name; and in Boston, David Page, a newly qualified MD and protégé of Bodmer's RFLP rival David Botstein, was just starting his own lab at the newly formed Whitehead Institute.

Goodfellow and Page are an interesting pair, about as different in personality and appearance as could be imagined, although their origins have some similarities: Page was born in rural Pennsylvania, Goodfellow in the depths of the East Anglian Fens, and both were the first in their families to go to university, Page to Swarthmore, and Goodfellow to Bristol. After that, things begin to diverge.

David Page, 1986.
(Photograph courtesy of Cold Spring Harbor Laboratory Archives.)

In 1978, west of the Atlantic, Page enrolled in the joint Harvard/Massachusetts Institute of Technology medical degree programme, which encouraged a mix of research and clinical training, and for the next few years he meandered creatively between clinical medicine (spending some time at a hospital in Liberia, where he met his future wife, Elizabeth), and benchwork. Following the advice of David Baltimore, a fellow Swarthmore alumnus, who also happened to have won a Nobel prize in 1975 for his work in the burgeoning molecular biology field, Page hooked up with David Botstein. Botstein had just proposed his RFLP idea, and Page's assignment in his lab was to begin on the genetic linkage map that would eventually encompass the entire human genome, a rather ambitious project to give to a wet-behind-the-ears medical student getting some lab experience. However, it panned out for Page because the first RFLP that he found came from a region of similarity between the X and Y chromosomes; as Page says, his entire career has been determined by a toothpick, then the weapon of choice for picking clones from the agar jelly plates on which they were grown. Towards the end of his MD, with a pleasing circularity, Botstein recommended Page back to David Baltimore, as a possible inhabitant of the newly fledged Whitehead Institute, and Baltimore hired him in 1984, aged 28, as the first Whitehead Research Fellow.

Leaving David Page for the moment, standing amongst a pile of cardboard boxes in an empty laboratory and wondering what on earth he had got himself into, we return to Peter Goodfellow, who had an equally busy run-up to his date with the Y chromosome. In 1972, after a degree in microbiology from Bristol, Peter went to do a PhD with Walter Bodmer in Oxford, partly because he'd been advised to get into human genetics because it was the coming research area, but partly to show all of the teachers at his old school that, contrary to their assertions, working-class boys from the Fens could, indeed, make it into the Oxbridge system. The Bodmer lab was absolutely flying science-wise in those years, and Bodmer was a committed and inspirational PhD supervisor, so Peter became an expert in somatic cell genetics, publishing 13 papers during his PhD, five of which were in *Nature*. After that, he went off to Stanford for a postdoc. Outside the lab, his social life thrived; he was spat on by Sid Vicious at the last Sex Pistols concert (realising in the process that you don't necessarily need talent to be successful), worked for Jane Fonda and Tom Hayden during Hayden's bid for the U.S. Senate, and hung out with his new wife Julia and their friends in Berkeley and San Francisco. Unfortunately, life in the lab was not so good. Peter's project centred on the T/t complex, at that time believed to be the key controller of early mouse development, but as his work progressed, he realised he was in the awkward position of having to tell a lot of understandably sceptical scientists that they had got something important wrong—the T/t complex theory was a dud. Overturning the consensus in the field was unpleasant and time-consuming, although it did teach Peter how to function in adversity, a lesson that his glory days in Oxford had not required him to learn.

Peter's arrival at ICRF was something of a shock. Scruffily dressed, rake thin, trailing a cohort of lab members whose dress sense showcased the "punk meets new romantic" vibe of the time, he was a bit of a novelty to some of the crustier members of staff, to put it mildly. The wild-looking new lab head was in addition highly driven, mostly by an intense curiosity regarding science, but also by a fierce ambition to succeed. Combined with what Peter describes as a slight social autism, this drive meant that he was a ruthless adversary in a scientific argument, taking no prisoners and never tolerating mediocrity. However, these sometimes intimidating qualities masked a person entirely lacking in malice, with an almost palpable delight in having found something he loved on which to spend his working life. His scientific persona, combined with amongst other things his penchant for poetry (good) and football (perhaps less good, given his allegiance to the Arsenal team of the 1980s), made Peter a complex, brilliant man who was a delight to his friends and a red rag to his enemies.

Pleasingly, one of Peter's new colleagues in London, who was to provide him with much help, advice, and support in his future travails, was Guido Pontecorvo, who had been one of those to kickstart the field of human genetics all those years ago. Ponte had been lured down to the ICRF from his position as Professor of Genetics in Glasgow by Michael Stoker because he was tired of administrative chores (although as his filing system for official papers comprised a wastepaper bin in his office, the chores really can't have been that arduous). Mostly retired, he occupied a position at the ICRF best described as the Institute's scientific godfather, a generally benevolent presence, tempering the scientific cut and thrust with kindness. As well as providing an intelligent and knowledgeable ear for new ideas, he made one more major technical contribution to the somatic cell genetics field during his time at ICRF, showing that polyethylene glycol could be used to fuse cells together. This innovation, although apparently small, turned cell fusion

Peter Goodfellow, ca. 1987. *(Photograph courtesy of Viesturs Simanis.)*

into a reliably efficient process and became the method of choice for many years.

Despite the substantial presence in human genetics at the ICRF, Peter, partly because he didn't want to live forever in Walter Bodmer's shadow, had a very different idea for his future research. His years in California had got Peter very interested in developmental biology, and he returned to England with the intention of working on mouse development. However, circumstances conspired against him because the quarantine laws of the day meant that it would have taken 18 months to import all of the mouse strains he needed from the States, by which time he would be almost a third of the way through his contract at the ICRF. Given that the conversion of that temporary, tenure-track contract into a prized permanent tenured position depended on his getting some data fast, Peter made the prosaic decision that while he was thinking what to do next, he would keep on with what he was good at, make himself useful, and get some papers out quickly. He succeeded admirably. Whilst at Stanford, Peter had used his expertise with somatic cell hybrids to bring the brand-new technique of making monoclonal antibodies to the Bay Area, and this skill was also much in demand at the ICRF. His small lab published 31 papers during the next five years, mainly joint ventures, one of which has had lasting importance. His collaboration with Mike Waterfield (described in **Chapter 5**), during which Peter dreamt up an innovative way of making a monoclonal antibody against the human Epidermal Growth Factor Receptor, made possible the Waterfield lab's landmark discovery that the EGF receptor had been stolen by a virus to become the v-*erb-b* oncogene.

By 1984, this and his other successes meant that Peter had been given tenure, had expanded his lab, and was well on the way to becoming ICRF royalty. He had added molecular biology and the new DNA mapping techniques to his already formidable skill set, and, most satisfyingly, had found the subject that he wanted to work on: the Y chromosome.

Peter, like Pat Jacobs, and, indeed, the toothpick-wielding David Page, only became involved with Y-chromosome genetics by accident. The immunologist Andrew McMichael, a friend from his Stanford years, had asked him to characterise an antibody, 12E7, raised against a surface marker that was supposed to be found only on T cells, the white blood cells involved in recognition and clearance of infections. Far from being T-cell specific, the marker that 12E7 recognised, MIC2, turned out to be on pretty much everywhere. However, a dull project suddenly became more interesting when Peter's lab mapped the gene encoding MIC2 to the X chromosome and then found that this gene, *MIC2X*, had a partner, *MIC2Y*, on the Y. They cloned both genes, and in 1983, *MIC2Y* became the first-ever published gene with a functional product to be cloned from the Y. Having landed on the Y, Peter realised that it was a wonderful unexplored landscape, perfect for someone whose skills lay in genome mapping, and that hidden within the empty landscape was a valuable prize: the gene encoding the Testis Determining Factor. Peter was hooked.

Back in Boston, David Page was having a bit of a crisis of self-confidence. His position as a Research Fellow of the Whitehead Institute was entirely novel, as was the Institute itself. Jack Whitehead, its benefactor, had got together with Nobelist David Baltimore with the intention of producing a kind of artists' colony for scientists. They wanted to attract the best possible people and give them the resources and intellectual freedom to be as wildly creative as they wished. They were clearly onto a winner because they managed to recruit the five founding members of the Whitehead from the top echelons of molecular biology and genetics. Plunged into this boiling intellectual soup, Page, a newly qualified MD with only a couple of years' lab experience, was simultaneously exhilarated by the exalted company and terrified that they would dismiss him as an intellectual pygmy. To his further alarm, his expectations of being a kind of glorified postdoctoral fellow with Rudolf Jaenisch, a world leader in the new field of making genetically modified mice, were confounded when it became clear that he was expected to be working independently. Page had to perform a hasty mental regrouping and decided, rather like Peter Goodfellow a few years before, to stick with what he knew and was good at. In Botstein's lab, he had started a collaboration with Finnish geneticist Albert de la Chapelle—who in 1964 had described the first cases of sex-reversed XX males—to explore how RFLP probes could be used to study such disorders. It seemed natural to carry on with the collaboration and, in keeping with the ethos of the Whitehead, to aim high. Page set about finding a research assistant with a capacity for infinite hard work; he too, had decided to clone the Testis Determining Factor.

The Y chromosome looks a little like a skittle, with a short arm (called Yp), and a long arm (Yq) separated by a constriction called the centromere, the point at which the chromosome attaches to the cell spindle during cell division. From work from Pat Jacobs's lab in the 1960s, it was possible to infer the rough location of TDF, on the short arm Yp, because if this region gets accidentally stuck onto an X chromosome, it causes the development of XX males. Slightly narrowing the search area, TDF could not be at the very tip of Yp, because this was the pseudo-autosomal region, the part of the Y chromosome alike enough to the X to pair with it; genes located there could not be unique to the Y, as TDF had to be. This narrowed the region down a bit, but the leap from a piece of DNA some 10 million base pairs (10 Mb) long to finding a gene, perhaps a few thousand base pairs in length at most, was still huge. It would be a formidable technical challenge that perhaps only Page and Goodfellow, and one other lab, that of Jean Weissenbach in Paris, had the expertise, muscle, and ambition to contemplate realistically.

To make a map of something, the first things required are the mapping tools, and in the case of the Y chromosome, these were Y-specific DNA probes, able to recognise sequences on the Y when used in Southern blotting experiments. All of the labs working on the Y chromosome spent the early 1980s mining DNA libraries in various ingenious ways to amass such probes, so that by 1984, a fair number of Y-specific ones were known. The Weissenbach and Page labs, together with their clinical collaborators Bernard Noël and Albert de la Chapelle, joined forces to produce a low-resolution map of the Y, by seeing which of a set of Y-specific probes were recognised by DNA from sex-reversed patients. The principle, known as deletion mapping, was as follows: XX males were likely to be males because they were carrying chunks of Y-specific DNA in their genomes. Therefore, some or all of the Y-specific probes should see this on Southern blots of restriction-enzyme-digested genomic DNA. The places that the probes were binding could then be roughly mapped along the Y chromosome by the panoramic photo principle, by looking for their presence in the individual snapshots coming from each of the XX males. For example, if one restriction fragment from an XX male was recognised by probes 1 and 3, and a second fragment from another patient was seen by probes 2 and 3, the two fragments must be next to each other, with the overlap defined by probe 3. Of course, in these XX male patients, each snapshot, whether of a large chunk of the Y panorama, or a tiny part, must carry the part of the Y chromosome containing TDF; otherwise, the patients would not be male. Therefore, identifying the part of the panorama that always showed up regardless of which patient's snapshot was on view would give a rough location for TDF. In molecular terms, this simply meant that the probes that lit up DNA from the greatest number of XX males were closest to TDF.

In 1986, the two labs published the first deletion map of part of the Y chromosome, showing that TDF had to be in a region of the short arm of the Y that they called Interval 1. Interval 1 fulfilled all of the criteria above —it was there in all of the XX males with additional Y DNA, and furthermore, it was within a larger region missing in an XY female. Interval 1, however, was not small; it still covered ~3 Mb.

Peter Goodfellow's lab had also been mapping busily, but using different methods. Rather than relying on patient material, which was very hard to come by in London for various annoying nonscientific reasons related to the egos of the clinicians involved, they had gone for a combination of traditional and cutting-edge molecular biology techniques, based around the *MIC2Y* gene. *MIC2Y* and its partner *MIC2X* lie in the pseudo-autosomal regions of the sex chromosomes, meaning that *MIC2Y* was at the tip of Yp, in the region next door to where TDF would be found. By meiotic mapping, looking at how often pseudo-autosomal genes were able to switch between the X and Y chromosomes during the chromosomal do-se-do of meiotic recombination, the Goodfellow lab realised that *MIC2Y* hardly ever switched, in contrast to the rest of their pseudo-autosomal-specific markers, which changed chromosomes with happy abandon. This reluctance to switch placed *MIC2Y* very close to the boundary between the pseudo-autosomal region and the Y-specific region next door, and therefore made it a very good point of departure from which to start mapping the rest of Yp, and TDF.

Armed with the knowledge of *MIC2Y*'s location, Peter's graduate student and postdoc, Catrin Pritchard and Paul Goodfellow (a Canadian, unrelated to Peter), made a long-range restriction map of the short arm of the Y, using a new technique, pulse field gel electrophoresis, which is able to separate enormously long pieces of DNA on agarose gels. As a source of male DNA, they used a human cell line called Oxen, which had managed to acquire four Y chromosomes, making the task of seeing faint fragments on Southern blots much easier. The strategy was to use many different restriction enzymes to perform multiple digests of Oxen DNA, separate out the bands of digested DNA on gels, and then probe all of the different digests with four probes known to span the whole region of the Y short arm from *MIC2Y* right up to the far side of Page and Weissenbach's Interval 1. A panorama of Yp could then be assembled from the individual snapshots. This worked very well, and the study was published in *Nature* in 1987. Its appearance in such a high-profile journal was partly due to an additional observation made by the trio; at one particular point in Yp, there was a cluster of recognition sites for restriction enzymes whose cut sequences all contained the dinucleotide CG. Such clusters (called CpG islands) were the new hot item in molecular biology, because they had been shown by Adrian Bird and colleagues in Edinburgh to frequently mark the beginnings of genes. Could the Yp CpG island be marking the position of the gene for TDF, 250,000 bases from *MIC2Y*?

What do you do when you have a long way to go and no form of ready transport? You walk. By late 1985, the Page and Goodfellow labs both knew where they needed to be, but to get there was going to be a terrible slog. The routine of making DNA, digesting it, running it on gels, blotting the gels onto the membranes needed for Southern blots, making probes, hybridising the probes to the blots, and then interpreting the results is as boring as it sounds, and technically hard; getting the data clean enough to make any sense is a real challenge. However, the mapping performed in both labs up to now looked like chickenfeed when compared with what was to come, as now, they had to walk from the fixed points they knew into unknown territory. The Yp panoramic photos to date were complete, and correct, but they were like looking at the Earth from space; the resolution was far too low. What had to happen now was the zooming in, the molecular equivalent of Google Earth, to find the one spot occupied by TDF.

Chromosome walking in those days was extraordinarily tedious. You began with a probe a few thousand base pairs long from the known anchor point on the chromosome, which in Goodfellow's case was a clone from *MIC2Y*, and in Page's the first clone he'd ever isolated in Botstein's lab, which flanked the other end of Interval 1, towards the centromere. Using a library of Y-specific sequences, in which each library clone contained at most 30,000 bp of Y DNA, you'd hybridise the anchor probe to the library. Any clones lit up by the probe were picked and grown up to amplify the DNA they contained, and then the DNA was restriction-digested along with the anchor clone. You kept looking until you found a clone that had some digested bands in common with your anchor, and then, you had an overlap. After that, you started the process all over again, using as the new probe the region of your new clone that was farthest from the anchor clone. If you were lucky, you did this 50–100 times and ended up with a beautifully detailed panoramic photo of your region of interest.

The Page and Goodfellow labs were not lucky. During their previous low-resolution mapping adventures, it had become very clear that the Y chromosome was going to be a complete pain to analyse: Almost all of the probes isolated, instead of recognising one nice, clean band on a Southern blot corresponding to one locus, saw multiple bands; the Y was covered in repetitive sequences, some of which were also found on autosomal (nonsex chromosome) and X-chromosome DNA, and some of which cropped up several times on the Y itself. This was a mapping nightmare; if the overlap clone that you found was full of repetitive sequences, it was useless, because it could be from anywhere. In Peter's words: "We spent two years doing the longest walk ...: Clone, subclone, pull out a probe, find something repetitive, go back and start again, chunk by chunk by chunk. It took four to five people two years ... pretty brutal work."

In Page's lab, things were moving equally slowly, perhaps more so, because at that point the Page laboratory comprised just two people: David himself and his research assistant, Laura Brown, with occasional help from Harvard undergraduates with research project assignments (their Y genomic library was made by one of these, Jonathan Pollack, now a professor at Stanford). They evolved a system in which Laura worked the day shift and David worked nights, so the lab was a 24-hour operation. (Fortunately for his future happiness, David's wife-to-be was a dermatology resident in Toronto at that point, and thus remained blissfully ignorant of her boyfriend's Stakhanovite tendencies until she moved to Boston some time later.) David and Laura's task was made no easier by the fact that although David knew where to start, he didn't know which direction to take, and therefore had to try both ways. His probe was in a region of the Y with homology to the X, and only when he started to find sequences not hybridising to the X would he know that he was going the right way, into Yp. In the summer of 1986, he came into the lab for his night shift and realised from looking at Laura's data that they had finally walked into the Y-specific sequences, and knew which way to go. Before heading off for a rare holiday the next morning, he left a note for Laura, picking up on that summer's hottest MTV hit from Aerosmith/RunDMC; in an unlikely collision of heavy metal, rap, and science, Laura came in the next day to find a piece of paper with "WALK THIS WAY!" written on it.

Not surprisingly, as the two labs crept slowly towards their destinations through 1986 and 1987, the atmosphere was pretty intense. In human genetics, although the prize of cloning an important gene is very great, both in terms of potential utility to medicine and career advancement, the stakes are equally high. There are no consolation prizes if you get beaten, and nobody remembers the runner up. At the annual Y-chromosome conferences, the occasional dark horse popped up with some interesting data on TDF, but Page and Goodfellow were the star turns, their back-to-back presentations always the hotly anticipated event of the meeting. Living in such an atmosphere of excitement tinged with alarm is a pressurized existence, but it is in those moments that labs truly come alive, catching the wave rather than paddling aimlessly. The question, however, was who would reach the shore first.

By late 1986, Page and Laura Brown had narrowed down the part of Interval 1 carrying TDF to Interval 1A, adjoining the pseudo-autosomal tip of Yp; by studying more sex-reversed XX male DNA samples, they had found that this was the smallest part of the Y chromosome that could still specify maleness. Crucially, they also had a very interesting XY female DNA sample, WHT1013. WHT1013 had the misfortune to have had her father's Y chromosome accidentally mixed up with autosomal chromosome 22 (formally, a reciprocal Y;22 translocation), meaning that she had apparently acquired the whole of Yp except for Interval 1A2, a subregion of Interval 1A, and a region next door to it, Interval 1B, which was known from the XX males not to be necessary for maleness. Interval 1A2 was 140,000 bp long, and all of the data suggested that TDF had to be in it somewhere.

The reduction in size of the region to be searched now meant that Page could start looking for TDF by other means, as well as the laborious chromosome walk. One of the criteria for TDF was that because it was such an important gene, it was likely to be very well conserved throughout evolution; the TDF gene in humans should be extremely similar in sequence to the TDF gene in rats, cows, dogs, and other assorted mammals. Therefore, Page wanted to start doing some Noah's Ark blots, in which male and female DNA from multiple species could be restriction-digested and tested to see whether it would stick to any of the probes derived from the chromosome walk. In early 1987, a second Harvard undergraduate, Becky Mosher, started in the lab. Despite her complete lack of laboratory experience, she turned out to have a natural aptitude for molecular biology, and so David put her onto the Noah's Ark project. In April 1987, returning from a meeting in Oxford, Page was met by Becky in a state of high excitement—one of her probes had detected a band on the Noah's

Ark blot that was present in the male but not the female of every species on there. Once the DNA sequence of the probe was determined, it was clear that there was an open reading frame, a sequence from which protein could be translated. Even better, the protein encoded had a recognisable sequence motif, a Zinc Finger, which cropped up in transcription factors, proteins dedicated to copying the DNA template into RNA. Finally, the Page lab had found a perfect candidate for TDF and were even in a position to speculate about what it did.

The point at which you know you have something really important, but you still have to do the confirmatory experiments and then get the paper published before you get scooped, is just excruciating. Nothing can ever be done fast enough, nothing can allay the gnawing anxiety that even as you slog away, trying to be careful but trying to be quick, someone else is doing the same work, is maybe a little closer than you.... You don't sleep properly, you don't do anything other than go to work, you neglect your nearest and dearest, and the whole time, a little kernel of panic in your gut nags away like a physical pain. Therefore, it was hardly a surprise that the Page lab went into overdrive. Fortunately for David, the spring of 1987 had seen the arrival of his first tranche of postdocs, all attracted by his growing reputation in the field. They all set to work with a vengeance, blotting, mapping, sequencing, and checking that everything was right, that there was no possibility of error. And they got there. In November, they sent a paper to *Cell* entitled, "The Sex-Determining Region of the Human Y Chromosome Encodes a Finger Protein," and on Christmas Eve, 1987, the paper was published.

In London on the morning of 24 December, Gerard Evan's friend and colleague Mike Owen, still working in the ICRF unit at St. Bartholomew's Hospital, was winding down his experiments for the Christmas break when he received a phone call from the main laboratories in Lincoln's Inn Fields. It was a postdoc from the Goodfellow lab, and he sounded very worried. Would Mike come over straight away and talk to his friend? Mike, who had known Peter since they were both graduate students, had seen Page's paper, and left immediately. He found Peter underneath his desk, distraught, and spent the next three hours trying to persuade him to come out. It transpired that Peter had had almost no warning of Page's coup, because Page had simply sent him a fax of the front page of the proofs the previous evening. To make matters worse, Peter had been right on the verge of cloning the new gene, called *ZFY* by Page. *ZFY* turned out to map right next door to the CpG island on which the Goodfellow lab had recently published, and all of the lab's efforts over the last few months had been going into cloning it.

After such an event, it takes a while for a lab to pick itself up, dust itself down, and think what to do next, and the Goodfellow lab was no exception. Peter spent the next few months half-heartedly doing experiments, reading a lot of poetry, and drinking too much coffee. His postdoc Paul Goodfellow did manage to cheer him up slightly by pointing out to him that there were an awful lot of people in the world who would like to be failures like them, sitting in a well-funded lab full of highly talented scientists, but moving on to the next thing was always going to be hard. Matters were made no easier by the world's press; with alarming eagerness, they had latched onto the story of how one gene can make the difference between becoming male or staying female, meaning that Peter had to dole out quote after quote saying how important the work was. Even worse, scenting a great human interest story, the press reframed the whole story as a big transatlantic fight to the finish, won by the quiet American, lost by the flamboyant Englishman, completely ignoring the tedious reality that science and, indeed, life is never that simple. Early 1988 was not a good time.

Curiously, David Page was not having quite as much fun as he should have been either. With the announcement of the cloning of *ZFY*, he had been catapulted from being a medium-successful scientist with a growing reputation amongst his immediate circle into a media phenomenon. He was on the NBC News, the front page of *The New York Times*, and many other papers worldwide, and even made it into a local paper in the Caribbean, much to the bemusement of his postdoc Lizzy Fisher, taking a well-earned Christmas holiday there in the hope that she could keep *ZFY* out of her life for a short while. He didn't like it at all. The nature of the attention meant that some of his scientific colleagues, whose opinions of him had been fairly neutral up to this point, felt obliged to become more extreme; although his close colleagues were delighted at his success, many others prone to jealousy began to snipe at him.

Worse still was the attention from outside science. Page's public visibility changed instantaneously from almost zero to maximal, and it took years for the spotlight to wander away in search of other prey. In addition to receiving heart-rending letters from people hoping he could somehow solve their reproductive and sexual problems, he rapidly became a whipping boy for those on the loonier fringes of feminism, gender studies, and psychology. He was accused of sexist thinking: Why did he talk about the "dominant" Testis Determining Factor? Why was he categorising the female state as passive, and the male state as active? Why did he seek to oppress the bisexuals, the homosexuals, the sexually ambivalent, by his assertion that sexuality was bimodal and determined by a single gene? It was Page's misfortune that what the Testis Determining Factor did was very easy to grasp and equally easy to misinterpret.

Back in the relative sanity of the laboratories, investigators in the sex determination field were reformatting their ideas to take into account Page's new information and beginning to climb the next experimental mountain. Contrary to what the world's press had decided, *ZFY* was still only a candidate for the Testis Determining Factor; in human genetics, a gene cloned by reverse genetics can only be definitively assigned to a condition when it has passed all reasonable scientific tests, and *ZFY* now had to take those tests. The first, that it should be in the sex-determining part of the Y chromosome, was clearly fine, and that it appeared to be the only gene in a rather empty area of DNA was also in its favour. Whilst labs around the world got hold of *ZFY* probes and tested them on their own sex-reversed patient samples to make sure that the equations *ZFY* = male, no *ZFY* = female held in all cases, other experiments also got under way, but this time not in humans. Sex determination as a problem of developmental biology had long been of interest, and, unsurprisingly, the mouse model people were very interested in checking out *ZFY* in their favourite furry organism.

At this point, a figure in London hitherto in the shadows needs to be introduced; Robin Lovell-Badge, urbane, charming, clean shaven, immaculately dressed in his trademark jeans and ironed white shirt, was the complete opposite of Peter Goodfellow, but their collaboration, brokered by Anne McLaren—nemesis of the HY antigen theory of sex determination and Robin's boss at the MRC Developmental Biology Unit at University College London—was to prove immensely successful. Robin's technical expertise lay in mouse embryology, specifically in working with embryonic stem (ES) cells, and in making transgenic mice by microinjecting foreign DNA into fertilised eggs, and Anne had hired him in the hope that she could persuade him to work on her own field of sex determination. After the debacle of the HY antigen and other similarly misguided ideas, mouse sex determination was in great need of some new approaches. Anne correctly realised that a partnership between an outstandingly good human molecular geneticist and an outstandingly good mouse developmental biologist would allow a synergistic exchange of ideas and technology that would be of huge mutual benefit.

Based on the idea that TDF in humans and mice was likely to be conserved, Robin and Peter decided in 1985 at the start of their collaboration that a good strategy to clone TDF might be to try to express the human gene in mice, and hope to change female mice into males. Unfortunately, the idea was so far ahead of its time that it failed; to get the mystery gene into mice involved putting enormous chunks of the human Y chromosome into ES cells, selecting the cells that were expressing the MIC2Y antigen, and then making mice with the MIC2Y-positive cells. Because virtually every single technique in the process had to be made up as they went along, there were just too many obstacles to success.

Undaunted, Robin got talking to Liz Robertson, another transgenic/ES cell pioneer, and came up with another idea, to mutate mouse *Tdy* (by convention, and as *not* detailed in the previous chapter, gene names in humans are all capitalised, and those in mice appear as proper nouns; to further confuse matters, "TDF" is referred to as "Tdy" in mice) by hopping retroviral DNA into mouse ES cell genomes. If the retrovirus hopped into the *Tdy* coding sequence, it would disrupt the ability of the gene to encode *Tdy* protein, and therefore any mouse made from the mutant ES cell would be female irrespective of its genetic makeup. Slightly amazingly, this idea actually worked, giving rise to an XY female mouse that was even able to breed, albeit with huge difficulty. Unfortunately, when the DNA from this animal was analysed, Robin and Liz were unable to find any trace of retroviral DNA, which they needed as a flag to mark the position of the *Tdy* gene, so they were left in the teeth-gnashingly frustrating position of having mutated the *Tdy* gene but being no further on towards finding out what it was.

This new annoying mouse mutant, christened Tdy^{m1}, joined two existing sex-reversed mouse lines: *Sxra*, in which a small chunk of the mouse Y chromosome was stuck onto the end of the X, leading to XX male animals, and *Sxrb*, in which the same process had happened, but a smaller piece of the Y was transposed. Page's paper had shown that in mice, there were two *Zfy* genes, and that whereas *Sxra* carried both, *Sxrb* only carried one, *Zfy2*. Robin and Peter both realised that despite their misery at being scooped, they would at least be able to look at mouse *Zfy* in these animals, and that furthermore, because an awful lot more was known regarding the earliest stages of mouse sexual development than human, they could check whether *Zfy* was switched on in the right places and at the right time to be involved in sex determination.

What Robin's lab found regarding mouse *Zfy* was rather unexpected. There were actually four, not two, copies of *Zfy* in the mouse genome, two on the Y, one on the X, and one on an autosome, making it very hard to see how the gene could be functioning as a Y-chromosome-specific sex-determining factor, unless the prevailing ideas regarding how sex was determined were radically wrong. Because this was actually perfectly possible, given how little was known regarding how *Zfy* might work, this did not constitute solid evidence for the prosecution, but the lab's next finding was a bit more worrying for *Zfy* supporters. However hard the lab tried, they were unable to find any expression at all of *Zfy2* in foetal gonads, only in adult testes, and although *Zfy1* was detected, it was only on in the germ cells, which play no part in testis determination; in the crucial somatic cells of the foetal gonad, there was no sign of either gene. The clincher consigning *Zfy* being mouse *Tdy* to the bin was that in Robin and Liz's Tdy^{m1} sex-reversed XY females, there was no change in any of the four *ZFY*-like genes—they were all there, and on at normal levels. There was absolutely no way they could be responsible for maleness.

These findings were so startling that although they got their first inkling of what was going on quite quickly after the Page *Cell* paper was published, Robin's lab sat on the data for some considerable time, in order to make absolutely sure that they were correct. In the meantime, however, other doubts were starting to surface regarding *ZFY*. One of the earliest came from an unlikely source: a marsupial genetics lab in Australia, run by Jenny Graves. Page and Goodfellow both contacted her independently to get her to probe the marsupial genome, to check that marsupial *Zfy* was also on the Y chromosome. To everyone's surprise, it was not; it lay on an autosome, which was very unexpected, because maleness in marsupials is also specified by the presence of a Y chromosome. Although the discovery made it onto the cover of *Nature* in late 1988, in an unlikely collaboration between all three labs, it was still not conclusive proof, because marsupial sex determination could quite feasibly just be a bit different from the mammalian mechanism. At this point, Page, certainly, was not terribly worried.

So what was going on in the Goodfellow lab amidst this hive of activity of collaborators and competitors? Peter, after his first few months of despondency, had woken up to the possibility that David Page might have got it wrong. The catalyst was a crucial conversation with Paul Burgoyne, who worked on mouse sex determination at the NIMR in Mill Hill. Burgoyne had been to a meeting at which he'd seen a poster from Marc Fellous of the Institut Pasteur in Paris, describing some interesting human sex-reversal pedigrees from North African families. He realised to his excitement that the region of the Y chromosome implicated in sex reversal in these patients was not the region containing *ZFY*. Instead, it had to be lying close to the pseudo-autosomal boundary, almost at the limit of the Y-specific region on Yp. Once he got back to London, he went to see Peter. Between them, Burgoyne and the data were so persuasive that Peter could not help but be convinced to take another look at the pseudo-autosomal boundary region. He was back on the case, and if Page had by any chance made a mistake, he was determined to find it.

Peter contacted Marc Fellous, an old friend who had been a postdoc with Walter Bodmer when Peter was a student, and Fellous sent him his patient DNA samples. The four test samples were positive for the Y pseudo-autosomal boundary region, which Peter's lab had recently cloned, meaning that the sex reversal was likely caused by an abnormal exchange between the X and Y chromosomes and should therefore be due to the presence of TDF, and hence *ZFY*. However, as Burgoyne had foreseen, all tests to find *ZFY* came up negative, and further mapping of exactly which part of Yp was present in the patients showed conclusively that all four of the chromosomal breakpoints were far closer to the pseudo-autosomal tip, ~60,000 bp away from the boundary, and nowhere near Interval 1A2, where *ZFY* was located. *ZFY* could not be TDF.

ZFY could not be TDF. Peter describes the lab's reaction to this incredible, mind-blowing news as "pretty excited," which may be a little bit of an understatement. Combined with the Lovell-Badge data, it meant the end of *ZFY*'s short time in the limelight. The back-to-back papers describing the downfall of *ZFY* were published in the 21 December 1989 issue of *Nature*, two years after Page's *Cell* paper; working on sex determination seems to bring with it the hazard of ruined Christmases for one's competitors.

Once the celebrations in London were over, Peter and Robin's labs returned to the drudgery of sifting through the horribly convoluted maze of the Y chromosome, looking for the right gene. On their first walk through the boundary region of Yp, on their way towards the CpG island next door to *ZFY*, nothing particularly obvious had popped up, except that the region was stuffed full of repetitive sequences; it was clear that whatever TDY was, it was going to be fairly nondescript. Further mapping experiments using clinical samples from France narrowed the region to be searched down to the 35,000 bp abutting the pseudo-autosomal boundary, and thus the two labs decided to make probes by chopping this region into small bits. After eliminating the repeat sequences, the plan was to see whether any of the remaining probes could detect areas of homology in other organisms. Andrew Sinclair, a newly arrived Goodfellow postdoc from Jenny Graves's marsupial lab in Melbourne, began probing Noah's Ark blots. John Gubbay in Robin's lab used a slightly different approach, hybridising his probe set to blots containing DNA from normal male and female mice, and also *Sxrb* and Tdy^{m1} sex-reversed females. Gubbay and Sinclair both hit pay dirt at almost exactly the same time: Sinclair picked up a sequence that was conserved in multiple mammalian species, but only in the males, and Gubbay saw a band that was present in normal male mice but absent in both normal and sex-reversed *Sxrb* and Tdy^{m1} females. Fortunately for the sanity of those concerned, the probe responsible was the same in both cases, corresponding to a region with an open reading frame able to code for a short protein that, like *ZFY*, had all the characteristics of a transcription factor. By what is almost certainly an enormous coincidence, a similar protein had already been shown to control mating in yeast. The human gene, which Robin and Peter called *SRY* (for sex-determining region Y), had been missed before because it was tiny, only 896 bp long, and had no CpG island associated with it to give away its location.

In July 1990, the two labs published their findings in two articles in *Nature*. Together, they had constructed an almost watertight case. The Goodfellow lab had done the human genetics, the positional cloning, and the sequencing, and had further shown that *SRY* was conserved in the males of multiple species, and in humans was only on in testes. Lovell-Badge's lab had cloned the equivalent mouse gene, shown that it was male specific in mice, and that, crucially, it was expressed during development of the gonads in the cells required for testis formation.

In the same issue of *Nature*, the Page lab published a paper in which the reason that they had been misled by *ZFY* was revealed. It turned out that the genome of XY female WHT1013, on which the identification of *ZFY* had been based, contained one extra surprise that had not been detected; instead of one deletion removing *ZFY*, there were two, and the second deletion covered the region in which *SRY* was found. Such an occurrence was almost unprecedented, and Page and his colleagues, on the evidence they had at the time, were entirely right to have published what subsequently turned out to be incorrect—Peter always maintained that had he reached *ZFY* first, he would have published exactly the same paper.

Although *SRY* was universally agreed to be an excellent candidate for the elusive TDF, there was still one final test to be performed: An experiment was needed to show that by itself, *SRY* could cause sex reversal. Robin, the transgenic mouse expert, set about introducing Sry into female mice and seeing if it would be enough to turn them into males. The technical side of this was pretty much a doddle for Robin's lab, but unfortunately, there had been an outbreak of Mouse Hepatitis Virus in the NIMR animal house (Robin had moved to NIMR from University College in 1988), and in order to eradicate it, there was a complete ban on breeding for some months. Robin and his lab kicked their heels and waited out the delay and, finally, managed to make five mice carrying multiple copies of artificially introduced transgenic *Sry*. One of these mice, very happily for all, had two X chromosomes and no Y, but was very clearly a boy, with all the right bits both inside and out. Randy, as he was christened, was very keen on mating with females but was infertile.

Robin and Peter wrote up the paper and sent it off to *Nature*, where it was duly accepted and scheduled for publication. Much to their delight, Robin's photo of Randy climbing on a bar with one of his normal male companions, proudly showing off his very male private parts, was accepted as the front cover image for the 9 May 1991 issue of the magazine.

The weekend before the paper was published, Robin had a few phone calls from journalists and thought that, as for the 1990 back-to-back papers, that would be it for publicity. Nonchalantly strolling into work on the day of publication, he still thought he would be having a peaceful day, perhaps drinking a little champagne with the lab. However, the news had gone viral. There was something viscerally exciting about being able to change the sex of a mouse by switching on one small gene, and Randy and his anonymous friend were splattered all over the newspapers; the *Independent* gave them half the front page. As had David Page before them, Robin and Peter were deluged with interview requests, but surprisingly, they attracted far less flak than Page, and the experience was less alarming. Robin recalls the only slightly awkward part was having to explain that the *Nature* cover star, far from relaxing in a luxury cage in the animal house, was, in fact, the Late Randy, because to check that

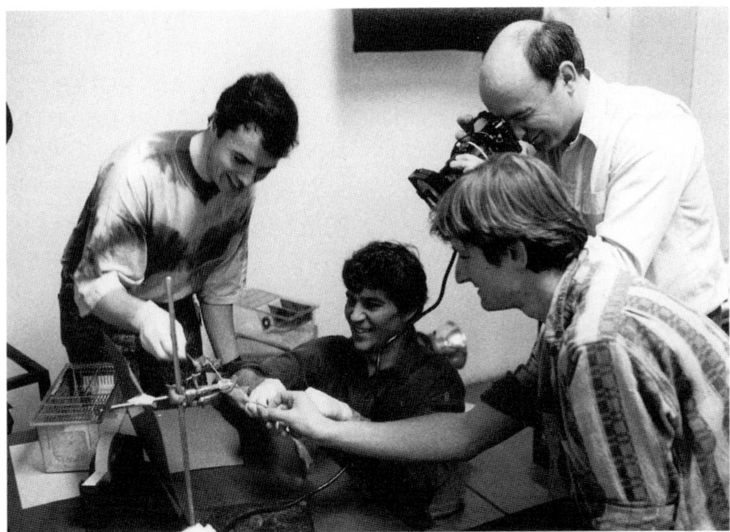

Photographing Randy. Robin Lovell-Badge is kneeling, centre.
(Photograph by Jérôme Collignon, courtesy of Robin Lovell-Badge.)

he was definitely male, he had had to be dissected. However, by way of consolation, one of his sex-reversed compatriots sits in the Science Museum, stuffed, beside Dolly the Sheep, and his fame persists to this day.

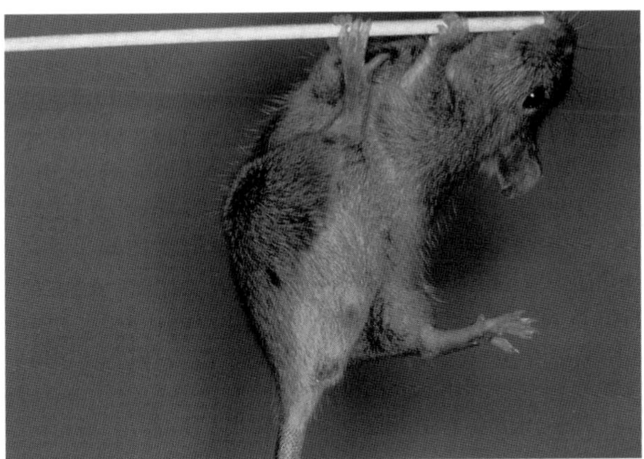

Randy. (*Figure used with permission of the Medical Research Council/Photo Researchers.*)

Aftermath

The paths of the three main protagonists of this story have diverged considerably in the 20-something years since Randy won the day for *SRY*. In the first months after *ZFY*'s downfall, David Page had to consider what to do next, and after much thought, decided, against the massed advice of his friends and colleagues, to carry on with the Y chromosome. His tenacity, superb bench skills, and infinite appetite for wringing the last drop of information from a complicated data set seemed to be perfectly suited to navigating the convolutions and complexity of the Y, and he refused to believe that the Y contained nothing more of interest. He was right: In 1992, his lab published a paper detailing the complete cloning of the Y chromosome, making it one of the first two chromosomes to be cloned, and signaling the renaissance of the Y. He collaborated in the sequencing of the Y chromosome in the late 1990s, finally resolving all of the hideous repeat sequences that had dogged the early chromosome walks and showing that they were there for a reason; the Y chromosome, uniquely amongst its fellows, uses the massive repeats to recombine with itself, maintaining genetic diversity. Since that time, Page has shown that the Y carries genes of paramount importance in spermatogenesis, and hence male fertility, work of great clinical significance. He is now Director of the Whitehead Institute, still in the same laboratory he first occupied as a 28-year-old novice. Looking back on the *ZFY* story now, Page is almost nostalgic. At the time, he says, it seemed perfectly fine to be working like a dog, maintaining his position at the front of the field by sheer determination and lack of sleep, and fighting intellectual battles with Peter Goodfellow at their regular conference encounters, because if you love something enough, that is what you do. The press attention and the trauma of the very public downfall of *ZFY* were horrible, but, despite all of his subsequent successes (and there have been many), the exhilarating rollercoaster of those days persists as the most exciting time of his career.

For Peter Goodfellow, now a biotech consultant after a high-pressure career in the upper echelons of the pharmaceutical industry, his years hunting down *SRY* also seem to have been a worthwhile adventure. The race for TDF was not just the big competition that the press liked to highlight, but was the result of many people's desire to find the same information. Although there was competition, there was also collaboration—sharing of probes, information, and patient material. Like David Page, Peter was bruised by having to endure all of his tribulations in the public eye, but finding *SRY* remains his proudest achievement, his "textbook fact," the major contribution to scientific knowledge to which every scientist aspires. He and Page, in the end, are more similar than they might appear; with Peter, as for Page, what comes out most strongly is his love of science, the vocational drive that motivated him to go for the big prize because it was what mattered most, not because it would boost his career and make him famous.

After the splash of the *SRY* papers, Peter started to get a lot of tempting job offers (although one of these was hastily withdrawn after he gave his job seminar sitting cross-legged on a table waving a pointer around), and in the end he succumbed. In 1992, the flamboyance quotient of the ICRF was reduced to a depressingly normal level when Peter moved to the University of Cambridge to become the Balfour Professor of Genetics. He still carries the distinction of being the only head of the Cambridge University Genetics Department to have a ponytail. In addition to radically reorganising the department, and reading poetry to the undergraduates, he continued to work on *SRY* and sex determination. *SRY* proved to be a difficult protein to study. After many years of effort, we now know that

Peter Goodfellow and grandchildren, 2013.
(Photograph courtesy of Peter Goodfellow.)

it is a rather weedy transcription factor, whose sole purpose as a testis-determining factor is to switch on a second gene, *SOX9*, cloned by Peter's lab in Cambridge. *SOX9* then does all the rest of the work in establishing maleness. Interestingly, *SOX9* is not on the sex chromosomes but is an autosomal gene, and appears to be a much more ancient specifier of maleness than *SRY*; it exists in multiple nonmammalian species, in contrast to *SRY*, which is solely mammalian.

And Robin Lovell-Badge? Robin's encounter with the media following Randy's appearance has morphed into a long-term relationship, and he is now a highly respected commentator and communicator on science issues, especially in the moral minefield of stem cell research. In the lab, he still works at the bench when he can, and his record of publishing high-impact work in high-profile journals is undiminished. In a recent twist to the story of *SRY*, the long-held dogma that femaleness was a default state, and to be male one simply had to activate *SRY* and *SOX9*, was overturned in a 2009 *Cell* paper, in which Robin and collaborators showed in mice that the autosomal gene *FoxL2* specifies femaleness, and when present, overrides *Sox9* and prevents ovaries changing into testes. Loss of *FoxL2* in adult female mice up-regulates *Sox9*, causing reprogramming of some ovarian cell types to those found in testes. Remarkably, as in life, the *Sox9/FoxL2* story shows that maleness and femaleness appear to be established by what some might view as a balancing act, and others as a war.

Web Resources

http://bsdb.satsumaweb.co.uk/2010/07/17/bsdb-newsletter-summer-2010 Profile of Robin Lovell-Badge, by his friend and collaborator Liz Robertson.

http://fds.oup.com/www.oup.co.uk/pdf/0-19-829792-0.pdf Peter Goodfellow discusses sex and molecular biology.

www.scivee.tv/node/11701 David Page talks about the Y chromosome.

Further Reading

Cohen BL. 2007. Guido Pontecorvo ("Ponte"): A centenary memoir. *Genetics* 177: 1439–1444.

Sulston JE, Ferry G. 2003. *The common thread: Science, politics, ethics and the human genome.* Corgi Books, London.

CHAPTER 9

The Hedgehog Three

In biology, everything is connected, sometimes by a far shorter link than anyone can imagine. This chapter tells the story of how work on an obscure mutation in fruit flies has led to the first drug treatment for advanced basal cell carcinoma. The plot takes in a grand German *Schloss*, an assortment of flies, fish, mice, and chickens, three extremely ambitious scientists, and somebody's very astute dad. But, first, let's start at the beginning, with the flies.

Drosophila melanogaster, the fruit fly, is responsible for spoiling many a summer picnic; few things are guaranteed to shatter a mood of lazy contentment more thoroughly than swallowing a fruit fly plundering the strawberries or bathing in one's glass of Pimm's. In laboratories, however, the fruit fly is a superstar, the organism in which the first steps were taken towards answering the million-dollar question of developmental biology: How on earth does a fertilised egg, a fairly simple-looking thing, grow into a complex organism?

Drosophila was the place where modern genetics began, in the Columbia fly lab of Thomas Hunt Morgan at the beginning of the last century. Morgan (who was the great grandson of the chap who composed the *Star Spangled Banner*; an entirely irrelevant but irresistible snippet) started using flies to explore Mendel's newly rediscovered theory of heredity because they had a short breeding time of 10 days and were very cheap to keep. After two years of getting nowhere, in 1910 he isolated the first fly mutant, a male with white instead of red eyes, and shortly afterwards showed that the *white* gene lay on the X chromosome. In subsequent work with further mutants, he confirmed the chromosomal theory of heredity, that genes lie along the chromosomes like beads on a string, and that different genes can be shown to be linked by tracking how often they segregate together during breeding. His student Alfred Sturtevant created the first genetic linkage map, transforming genetics from a descriptive, phenomenological discipline to a mathematical one; after Sturtevant, genetic distance could be mapped precisely and the order and distance apart of genes on chromosomes determined (a slightly more relevant snippet: Sturtevant was colour blind and must have had some difficulty distinguishing different fly eye colours).

In 1933, the year that Morgan won the first fly Nobel Prize in Physiology or Medicine, *Drosophila* received a further boost with the rediscovery of polytene chromosomes in their salivary glands. Polytene chromosomes form when a cell duplicates its DNA many times without dividing, leading to many (1024 in the case of *Drosophila*) copies of the four chromosomes. Each chromosome hangs onto its replicated sisters, forming enormous, thick ropes of DNA large enough to be seen in detail down a light microscope. The chromosomes each have their own unique stripy pattern of bands, rather like a badly knitted Dr. Who scarf, and in 1934, Calvin Bridges, another Morgan alumnus, produced a set of beautifully detailed drawings of the banding pattern, which are still used today to navigate around the chromosomes. Once the bands were documented, it became very obvious that polytene chromosomes provided a link between the fly geneticists' theoretical ideas regarding genes and linkage, and physical reality; a phenotypic mutation such as the red-to-white eye colour change could be correlated with a change in the banding pattern on a particular chromosome, thereby pinpointing the location of the mutated gene.

Aside from their cheapness and their handy polytene chromosomes, the other big advantage of *Drosophila* as a model organism was the ease with which mutations in the adult body plan could be spotted. Compared with complex vertebrate anatomy, there is not much to a fly, and mutations in legs, wings, bristles, and so on were visible with the aid of even the most primitive microscope. Professionals and interested amateurs alike could study a miniature freak show whose exhibits included flies with legs instead of antennae, flies with extra sets of wings or no wings at all, and flies with no eyes. Flies were the perfect organism for the hobby biologist; the only experimental work to be performed was setting up matings between different flies and then observing their offspring; all the rest of the work was the intellectual puzzle of determining which crosses to do and interpreting the results. Ed Lewis, fly geneticist extraordinaire, started working on *Drosophila* in 1935 as a schoolboy, buying stocks from an advert in the back of *Science* magazine and working after hours in the school science laboratories.

By the 1970s, fly labs had developed a repertoire of increasingly sophisticated genetic tricks, but their main experimental aims were still very similar to those of Morgan: mapping new genes and detecting new mutant versions (alleles) of known genes. However, the theoretical side of the field, suggesting the possible genetic mechanisms behind the mutant phenotypes, had flourished. The superficial answer to the question of how a single egg can turn itself into an organism was now clear; everything boiled down to which genes were expressed where, and when. However, it was frustratingly difficult to progress beyond this simple statement.

Drosophila are built on a segmented body plan that becomes apparent in the embryo during the first hours after egg fertilisation, when three head segments, three thoracic ("chest") segments, and eight abdominal segments are clearly visible. The thoracic and abdominal segments look pretty similar to begin with, but each ends up specifying a different part of the fly's anatomy. For example, the first thoracic segment eventually makes a pair of legs, the second a pair of legs and a pair of wings, and the third a pair of legs plus a pair of halteres (little knob-like balancing organs that the fly uses during flight). In the mid-1970s, the Spaniard Antonio Garcia-Bellido proposed that the strange mutant flies so beloved of the early *Drosophila* investigators could be explained by mutation or misexpression of "selector" genes whose function was to specify segment identity. For example, if a gene normally specifying a segment making legs was accidentally switched on in the head, legs would grow instead of antennae; similarly, extra sets of wings, or a complete lack of wings would arise when genes specifying particular abdominal or thoracic segments went wrong.

Working out how such selector genes caused these "homeotic" (from the Greek, *homoiosis*, "to make similar") transformations was a major preoccupation of the fly field, epitomised in 1978 by Ed Lewis's prescient *Nature* paper, which summarised and interpreted 30 years of his work on how a multigene complex regulated the wing-duplicating *bithorax* mutation. However, homeotic genes were not the only players in fly development. A whole set of other genes was involved in turning eggs into segmented embryos and larvae, but strangely, almost nobody seemed to be interested in them, perhaps because eggs and larvae are pretty boring to look at, and mutations affecting tiny wriggling tubes are rather hard to detect. All this changed in 1980, with the publication of a paper in *Nature* entitled "Mutations Affecting Segment Number and Polarity in

Eric Wieschaus and Janni Nüsslein-Volhard at Janni's 70th birthday party, 2012. *(Photograph courtesy of Jim Smith.)*

Drosophila," written by Christiane (Janni) Nüsslein-Volhard and Eric Wieschaus, then working at the European Molecular Biology Laboratory in Heidelberg.

Janni Nüsslein-Volhard and Eric Wieschaus's work arose from one of the great synergistic partnerships of modern biology. Janni, reserved, formidably smart, and extremely practical, had originally been a molecular biologist, and the more extrovert Eric, widely held to be slightly bonkers but brilliant, started as a pure geneticist. The pair first met in Walter Gehring's lab in Basel, one of the powerhouses of fly developmental biology and one of the few labs in the world whose inhabitants thought that the fly embryo mattered as much as the adult. From their work with Gehring, Janni and Eric realised that identifying the very earliest genes required for correct front-to-back (anterior–posterior) patterning of the fly embryo was crucial to understanding how the body plan was laid down, and that, moreover, they could use techniques lifted from bacterial molecular biology to find these genes. Their work solved the *which* issue of early development, providing the fly world with a complete list of the important genes and a hypothesis for how they worked. Moreover, their methods could be easily adapted to search for almost any developmentally important genes, not just in flies, but in other organisms. Together with

Ed Lewis's work on homeotic mutations, Janni and Eric's experiments ushered in the modern era of developmental biology, and their combined insights won all three scientists the 1995 Nobel prize for Physiology or Medicine.

Funnily enough, there was nothing in Janni and Eric's experiment that couldn't have been performed 40 years previously; it was just that they were the first people in the fly world who possessed the *nous* to realise what was important, the background in molecular biology to understand what was experimentally possible, and the ambition to try something entirely new and extremely labour-intensive—a saturation screen.

Saturation screens are rigged to find mutations in every single gene involved in a particular pathway, in this case, the pathway to a normal embryo. To get to saturation, you must make an informed guess regarding how many genes you might be trying to hit and then do the mutagenesis enough times to make the likelihood of catching everything as close to 100% as possible. With bacteria, which was where Janni got the original idea, this is pretty easy—you just need a week or so, a lot of agar plates on which to grow the bacteria, and some incubators set to 37°C. With flies, the principle is the same, but the logistics are a nightmare. With all of the genetic tricks necessary for a clean result omitted for clarity, Janni and Eric's experiment ran as follows: Mutate the sperm of a male fly by feeding him the chemical ethyl methanesulphonate, and mate him to a normal female; mate their mutant sons to more normal females; mate together the resulting grandchildren; look for dead embryos (one quarter of those born, because both parents have one mutant and one normal allele; remember Mendel's peas and Punnet squares), and analyse them for mutations in their segmentation pattern by staring down a stereoscopic microscope with your colleague and arguing whether what you can see is odd or not; repeat approximately 27,000 times, and additionally, don't forget to keep all of the 27,000 lines going, by transferring breeding stocks of the original flies once a fortnight into new vials (you need at least three vials per line). Once you have your mutants (which turned out to number 580), roughly categorise them according to what you see down the microscope, then cross all of the lines in each category together to work out which lines are mutated in the same gene. (If the mutants are in different genes, they will complement each other, restoring the normal phenotype so that all the embryos will be alive. If they are in the same gene, there is no complementation, and a quarter of the embryos will still die, as before.) Finally, think of a good name for your mutant, based on how it looks (whimsy is permitted).

Janni and Eric's experiment soon turned into Janni, Eric, Gerd Jürgens, and Hildegard Kluding's experiment, assisted by anyone competent who happened to be passing through (one ex-Gehring labmate recalls a weekend break to Heidelberg to visit his friends entirely spent looking down a microscope), but by 1980, it was clear that all of the work had paid off. Janni, Eric, and their coworkers had identified 139 genes that, when mutated, altered the embryonic segmentation pattern. The genes could be divided into three distinct groups, with each group affecting a particular aspect of segmentation. Group one, the "gap mutants," produced embryos in which multiple adjacent segments were absent. For example, *hunchback* mutants lacked two adjacent thoracic segments, leading to an embryo with a normal abdomen but almost no thorax, whereas *Krüppel* embryos had no thoracic segments and were missing the next five abdominal segments too. Group two, the "pair-rule mutants," produced embryos in which either odd- or even-numbered segments were absent or mutated. Group three, the "segment polarity mutants," had the correct number of 14 segments, but within each segment, either the front or the back section was deleted and replaced by a mirror image of the remaining part.

The three groups of mutants and their phenotypes allowed Janni and Eric to propose the ground rules for anterior–posterior patterning of the fly. Gap genes were the first to act, laying down the basic head-to-tail body plan, and this pattern was then worked on first by the pair-rule genes, and then by the segment polarity genes, resulting in the creation of segments. The idea was beautifully simple and very appealing by virtue of its common sense; like building a house, one started with the foundations, built the shell of the house, demarcated the rooms, and then put in the correct number of internal floors and walls. Furthermore, once it was known how the initial building work was performed, it became very clear which genes would do the genetic equivalent of the interior decoration of the empty house: the homeotic "selector" genes. Governed by the patterns of expression of the segmentation genes, and also by other genes of their own class, the homeotic genes specified segment identity, so that legs, wings, halteres, and so on grew in the correct places.

The *Nature* paper announcing the first results from the screen was viewed with interest by the fly field, and awe at the amount of labour involved by everybody else, but it wasn't immediately clear how important it was going to be. Part of the problem was that the experiment was slightly ahead of its time. Lists of genes were all very well, but the means of cloning those genes were not quite advanced enough, and many, including Janni herself, were not up for the labour involved. As she remarks in her Nobel autobiography, "In my lab, molecular analysis was begun rather late, as we felt it important to investigate the properties of the individual genes as carefully as possible before embarking on tedious molecular cloning, that was not easy at the time" (Nüsslein-Volhard 2012). It took three years until the first homeotic genes were cloned in the labs of Dave Hogness, Walter Gehring, and Thom Kaufman, and these were followed in short order by the first segmentation genes.

Once genes are cloned, a whole new world of experimental possibility opens up: One can see where in an organism the genes are expressed by looking at where their mRNA is to be found; work out their DNA sequences and from those, their protein sequences; see whether the proteins look like any others, and from that try to draw conclusions regarding possible function; make antibodies against the proteins and see where they are, both in the cell and in the body; and generally throw every weapon in the molecular biology arsenal at the problem. Although part of the fly world fought a determined rearguard action to prevent their elite, intellectual field from being trampled into the sticky mud of molecular reality, they were eventually forced to surrender, helpless in the face of a torrent of intriguing information provided by the new methodology.

Despite all of these exciting goings on, fly work and, indeed, the wider field of developmental biology, was still a decidedly minority interest in the scientific world. It was therefore a bit of a surprise that the ICRF, a cancer research institute, already had a thriving unit at its north London outpost in Mill Hill that was dedicated to studying the process of normal development in organisms as diverse as slime mould and frogs. As related in Chapter 4, the ailing ICRF Mill Hill laboratories had been rescued in 1973 by John Cairns, who turned their fortunes around dramatically. Cairns, in addition to establishing a group working on DNA repair and thereby laying the foundations for the incredible subsequent successes of the Clare Hall laboratories, believed passionately that to understand what went wrong in cancer, one had also to understand the process of normal development. To this end, he recruited Brigid Hogan, working on mouse teratocarcinomas, and Julian Gross, a slime mould person, who, before taking the slightly unusual career decision of leaving for

India to sit at the feet of the Bhagwan Shree Rajneesh, hired fellow slime fans Jeff Williams and Rob Kay. Jonathan Slack, a frog enthusiast, and David Ish-Horowicz, working on *Drosophila*, completed the developmental biology cohort in 1979.

David Ish-Horowicz, the *Drosophila* component at Mill Hill, was a biochemist who had been seduced by the possibilities of fly genetics. After a PhD at the Laboratory of Molecular Biology in Cambridge, he went as a postdoc to Walter Gehring's lab in Basel, overlapping with Janni and Eric, and becoming infected, like them, with an enthusiasm for the embryo. When he returned to London to set up his own lab at the ICRF, it was as one of the new breed of *Drosophila* molecular biologists, working out the peculiarities of the *Drosophila* genome and trying to clone and analyse fly genes. In 1982, inspired by Janni and Eric's work, he turned his attention to a segmentation mutant, the pair-rule gene *hairy*.

David was one of the first fly people to move into molecular analysis of embryonic genes, and *hairy* was a very good place to start. Unlike many segmentation mutants, *hairy* flies survive into adulthood; mutant flies are, as one might expect, extra hairy—they have excess bristles, the adult sensory organs, on parts of their wings, heads, and thoraxes. The adults are fertile, making it possible to breed them and apply some useful fly genetic tricks to pinpoint and then clone the *hairy* gene; this was far

David Ish-Horowicz and Jeff Williams.
(Photograph courtesy of Jim Smith.)

easier and faster than the awful slog of walking along the polytene chromosomes towards a mutant locus, then the only other alternative.

For this new project, because another postdoc was required to augment the existing inhabitants of the lab, David put an ad in *Nature* and crossed his fingers. Because he was a relative unknown, it would be hard to attract a person of the calibre he required. Fortunately, at the Institute of Molecular Genetics of Eukaryotes in Strasbourg, someone good, desperate, and immediately available was reading the job pages in *Nature* that week: Phil Ingham.

Perhaps out of a need to prove to the world at large that they are just as interesting as everybody else, scientists love embroidering the personal folklore of their peers. In the case of Phil Ingham, the rumour that he trained as a priest before becoming a scientist has dogged him for years. Sadly, as the thought of a young Phil clad in clerical black and preaching to a rapt congregation is one to relish, it's not true. He does have an A level in Religious Studies, to go with a clutch of science subjects, and he did do a year of Theology at Cambridge, having applied to do the Social and Political Science Tripos, but it was out of a desire to engage his brain with something more interesting than the dusty school science syllabus, rather than from any religious conviction.

Rebelling against science in favour of sociology was all very well, but Phil soon found that although the Theology course at Cambridge was rather good, the rest of the SPS Tripos was a fossilised relic and that there was no way he wanted to finish the course. This left him with the large problem of what to do instead. Given his A levels, Natural Sciences seemed the only option, and thus to give it a try, Phil wandered one day into a lecture on mitosis and fell in love. Having never done biology, he'd never seen anything like the wobbly black-and-white film showing the stately dance of the chromosomes lining up on the mitotic spindle, condensing and dividing in perfect synchrony, and he was entranced. It turned out, following a conversation with his Director of Studies, that it would be possible to switch and go directly into the second year of the Natural Sciences course, but the downside was that his options would be quite limited. Phil signed up to do History and Philosophy of Science (a no-brainer to someone who had already spent a year writing essays on subjects such as the impact of Darwinism on the psychology of religion), Biochemistry (allegedly easy if you'd done A-level chemistry), and Genetics, because it sounded interesting and he wasn't qualified for anything else.

The genetics course at Cambridge in the 1970s was fascinating, because the undergraduates were the first generation to learn not only classical genetics but also the new theories regarding *Drosophila* selector genes and homeotic mutations. Fortunately, because Antonio Garcia-Bellido's work was not an easy read, the students had Peter Lawrence, one of the best communicators in the fly business, to explain things properly to them. Developmental biology acquired an awful lot of extremely able scientists thanks to Lawrence's lucid and highly persuasive lectures, and Phil Ingham, hearing Lawrence one winter's evening, was one of them. To someone who would have loved to be a Cambridge mathematician, but (in his opinion at least) wasn't quite up to it, genetics, with its combination of maths and biology and the elegance of its logic, was very appealing, and fly genetics was just about perfect; the added dimension of being able to use genetic logic to draw conclusions regarding a subject as tantalising as development was unbeatable. With all traces of his sociology ambitions eradicated, Phil left Cambridge in 1977, bound for Sussex and a PhD in the laboratory of Bob Whittle, one of the few people doing fly work in the United Kingdom at that time. Once there, he thrived, isolating a new homeotic mutant, *trithorax*, whose most extreme phenotype was the development of flies with six wings.

In the summer of 1980, his final year as a PhD student, Phil went to a fly conference in Kolymbari, Crete, which, if he had been a little less naïve, he would never have had the nerve to apply to, let alone attend. Entitled "The Molecular and Developmental Biology of *Drosophila*," the meeting was a high-level get-together of the cream of the *Drosophila* world, and the chance of an unknown PhD student getting in would normally have been about as likely as a donkey competing in the Derby. However, Phil's *trithorax* work was cutting-edge stuff, and the organiser, Walter Gehring, was interested enough to get him along. Phil even got to give a talk, although he had to do it without slides because he hadn't thought to bring any with him. The experience wasn't entirely positive; after he'd finished, he was told with great relish by an unkind French scientist, who shall remain nameless, that the presentation had been a disaster.

The Kolymbari meeting was the first time Phil heard Janni and Eric talk about their saturation screen, just before it was published. The Whittle lab was entirely uninterested in embryos, so much so that during the course of his PhD, Phil had isolated some 2000 embryonic lethal mutations but had thrown them away without looking at them (a source of some regret after Janni and Eric won their Nobel). Phil managed to bag a seat between Janni and Eric at dinner one night, and although Janni was rather quiet, he was charmed by Eric, who talked away nonstop for the whole meal and converted him wholeheartedly to the embryonic view of things. Combined with the very molecular focus of the rest of the meeting, the existence of a list of segmentation genes that really needed to be cloned made Phil realise that to get any further with fly development, he was going to have to learn some molecular biology.

Doing a postdoc in Strasbourg with Pat Simpson, a renowned *Drosophila* person, seemed to be a good option. Pat's proximity to the many top-flight molecular biologists there should have meant that the techniques Phil needed to learn would seep through into her lab, but unfortunately, the molecular biologists seemed uninterested to the point of rudeness in any contact with fly people. In the end, the high point of Phil's time there was a trip to Tübingen, where Janni was now based and where their Trappist dinner in Crete was forgotten in a day of intense conversation that laid the foundations for a long and happy friendship.

Strasbourg was clearly going to be no use to Phil at all, so seeing David Ish-Horowicz's ad in *Nature* was a godsend. In a tribute to the power of conference networking, the two had already been introduced at the Kolymbari meeting, and thus, a year after his departure for Strasbourg, Phil found himself back in the United Kingdom again, but this time in a lab that was wholeheartedly of his own opinion that embryos mattered, and molecular biology mattered too.

The Mill Hill laboratories, although rather dilapidated, were a great place to work. The mix of different organisms there exposed Phil to the world of vertebrate developmental biology, and his colleagues were amongst the brightest and best then working in Britain. Phil found the intricacies of molecular biology, cutting and pasting bits of DNA together, to be intellectually quite similar to thinking through the breeding steps necessary to get correct fly crosses, but being a molecular biology novice in the Ish-Horowicz lab was fairly alarming. David's research assistant, Sheena Pinchin, a really good fly person, was fine, but at first sight, David's three existing postdocs and a brilliant PhD student, Ken Howard, were almost as scary as the uncooperative French scientists of Strasbourg. Fortunately, once they'd sized him up and realised his fly knowledge complemented their DNA skills, they rapidly transmuted into pleasanter, if still deeply competitive, beings. Phil particularly bonded with

Alfonso Martinez-Arias, Phil Ingham, and Ken Howard on the South Downs, 1986.
(Photograph courtesy of Phil Ingham.)

Ken Howard, David's PhD student, and the two of them, together with Sheena, got on with the *hairy* project.

In summer 1985, the annual Symposium at Cold Spring Harbor, pretty much the most prestigious meeting on the biology conference calendar, was devoted to the topic of the molecular biology of development. It was held in an atmosphere of high excitement, at a time when the first molecular truths of development were being revealed. Organisms from yeast through slime, worms, flies, frogs, and mice were all represented. People had made transgenic mice, their genomes altered by the heritable introduction of foreign DNA, and the first stem cells were being isolated. However, the fly talks took centre stage. Although only a handful of segmentation and homeotic genes had been cloned, the developmental insights they were already providing were pretty startling. Things had started to heat up the year before with the discovery of the homeobox by Bill McGinnis, Ernst Hafen, and Mike Levine in the Gehring lab and Matt Scott in Thom Kaufman's lab in Indiana, and the realisation that the homeobox-encoding DNA sequence was found in multiple fly homeotic and segmentation genes. By comparison with a similar sequence found in yeast, it was clear that homeobox domains could bind DNA, meaning that proteins containing them were likely to be transcription factors, able to control the expression of other genes, which was exactly what the products of major regulatory genes should be doing. This was exciting enough, but the reason that nobody was bunking off to the beach when the homeotic guys were on was that homeoboxes had also been found not only in yeast, but in vertebrates. Although there were as yet no functional data regarding the vertebrate homologues, their very existence hinted that developmental mechanisms might be startlingly well conserved, and that what held in flies might also apply in higher organisms. All present realised that these unexpected findings might be the way in to a previously intractable molecular problem, vertebrate embryology, and there was a tantalising possibility, subsequently borne out, that the vertebrate homeobox proteins were just the tip of the iceberg. Many more fly genes would have vertebrate counterparts.

Segmentation genes were also well represented at the meeting. On the programme, scheduled just before Janni, was a talk from the Ish-Horowicz lab on *hairy*, but unfortunately, its presenter was not in the best possible state to communicate anything other than a desire for antibiotics. Phil, completely stressed out by the whole rigmarole, had succumbed to tonsillitis and was stuck in bed in the sweaty confines of one of the old Cold Spring Harbor sleeping huts. As he recalls, "I was desperate to give the talk, as it was the hottest stuff I'd ever had, but for the first three days I just couldn't get out of bed. And Peter [Lawrence] kept coming to me every day with baskets of fruit, and Janni said 'Why is there this great love between you and Peter all of a sudden?' "

Fortunately, the invalid recovered sufficiently to give his talk on the last day. In the three years since Phil's arrival in David's lab, the *hairy* project had moved on a lot. David, with help from his postdocs Andy Leigh Brown and Julian Burke, had cloned the *hairy* locus, no mean feat in itself, but it was Phil and Ken Howard who had then gone on to define its expression pattern, together with another pair-rule gene, *fushi tarazu* ("not enough segments" in Japanese). To look at expression of the genes, Phil had turned himself into one of the very few *Drosophila* experts in in situ hybridization, a method for recognising and labelling mRNA in ultrathin sections sliced lengthwise through the fly embryos. Together with Ken, he had got some spectacular results. *hairy* mRNA initially came on all over the embryo, but its expression was rapidly refined into seven zebra stripes. *fushi tarazu* mRNA came on subsequently and also had a seven-stripe zebra pattern, but wherever *hairy* was on, *fushi tarazu* was off. The patterns for both genes corresponded to the segments of the embryo that were absent when the genes were mutated. Exactly how the refinement of the early widespread *hairy* expression into the subsequent stripes occurred and whether the *hairy* stripes defined where *fushi tarazu* came on was an intriguing puzzle that struck right at the heart of the patterning problem.

Phil and Ken's results, combined with work from the small number of other labs able to do in situs, showed that all of the segmentation genes cloned so far were expressed in distinctly different stripes, always corresponding to the parts of the embryo that were absent when they were mutated. Over the next year or so, as more labs got in on the act and provided further data, substantial flesh was put on the bones of Janni and Eric's original theory of segmentation. The gap genes came on first in broad stripes, in the part of the embryo that they governed. (Their pattern was preordained by maternally expressed genes, whose mRNAs are bequeathed to the egg by its mother.) Scanning down the embryo from top to bottom, there were now different combinations of maternal and gap proteins. The mix of proteins only persisted for a short time, but before their demise, they wrote personalised molecular messages to the pair-rule genes, telling each of them where to come on in their characteristic zebra pattern. Before they too faded out, the pair-rule genes specified 14 stripes of segment polarity genes, which added in the details that eventually resulted in a visibly segmented embryo. The 14 segments of the early embryo were half a segment out of register with the visible segments of the late embryo, and were therefore christened parasegments.

Phil's in situ and molecular biology skills, combined with his evidently superior *Drosophila* brain, meant that he was now extremely marketable, and by the time the *hairy* work appeared in *Cell* the following March, he was no longer a postdoc at Mill Hill. Peter Lawrence's solicitude at Cold Spring Harbor may have been a novel, fruit-based, recruitment technique, because he had lured Phil to the Laboratory of Molecular Biology in Cambridge, where, with two other stars of the new molecular fly field, he was supposed to spearhead the LMB's attempts to enter the modern age. Unfortunately, the plan unravelled rapidly with the nonappearance of his two promised colleagues, and Phil once again found himself unhappily marooned. Not many months in, he started looking for a means of escape. Salvation, when it came after a year of discontent, had a familiar face; once again, David Ish-Horowicz rode to Phil's rescue, although the stable from which he spurred his scientific horse was now in Oxford, not North London.

Following the decision to close the Mill Hill laboratories, the remaining developmental biologists there chose different escape routes, all of which proved to be remarkably successful: Jeff Williams, who needed to stay in London, went with the DNA repair people to Clare Hall; Brigid Hogan was recruited next door to the MRC National Institute for Medical Research; Rob Kay went to the LMB; and David Ish-Horowicz and Jonathan Slack moved to the newly established ICRF Developmental Biology Unit in Oxford, whose incoming director, Richard Gardner, was one of the big cheeses of mouse embryology.

The calibre of laboratory heads in the ICRF Developmental Biology Unit (DBU) was astonishingly high; most became Fellows of the Royal Society (Gardner already was one), and the haul of medals, institute directorships, and the other trappings of academic success they have racked up among them would fill a fair-sized trophy cabinet. On the top floor of the Zoology Department, David and Jonathan's new colleagues in addition to Gardner were Andy Copp, working on development of the mouse nervous system; Julian Lewis, a chicken person working on limb and inner ear development; and the extraordinarily gifted and charismatic mouse embryologist Rosa Beddington, still bitterly missed by her many friends and colleagues after her tragically early death in 2001 at the age of 45. Owing to Jeff Williams's decision to remain in London, there was still some empty space in the unit, and David, hearing of Phil's plight, engineered the return of his old postdoc into the ICRF fold, this time as an independent lab head.

Compared with the LMB, Phil remembers the DBU as a definite step up:

> It was well-supported; it was well-equipped. The labs had been refurbed—it had been the MRC molecular biophysics unit and the ICRF gutted it and did it up. I think David asked me if I would join basically for the same reason Peter Lawrence wanted me to go to the LMB—because I could do in situs. Space wise, it wasn't great, but compared with the LMB it was wonderful because I had a brand new Zeiss Axioplan microscope, I had my own dissecting microscope, I had my own space, and they gave me a technician and a student and a postdoc. I didn't have any of those at the LMB.

In their 1986 *Cell* paper, Phil and Ken Howard had not only shown that *hairy* negatively regulated *fushi tarazu*, but also that both genes had an effect on *engrailed*, a segment polarity gene. Upon his move to Cambridge, Phil had taken the *engrailed* project with him, not wanting to compete directly on pair-rule genes with his old lab, and soon hooked up with Nick Baker, a graduate student of Peter Lawrence's who had just cloned another segment polarity gene, *wingless*. Although both *engrailed* and *wingless* were initially regulated by pair-rule genes, it was clear that this was only a transient phase in the development of the embryo, and it was a mystery how they were controlled thereafter. Phil, Nick, and Alfonso Martinez-Arias showed that at subsequent stages, *wingless* and *engrailed* were expressed in adjacent rows of cells either side of the parasegment boundary, which they were responsible for maintaining, and that each was required for the expression of the other. Because Wingless was a nomad, a secreted protein, it could easily wander along to the next-door cell and tell it to switch *engrailed* on, but *engrailed*, a transcription factor confined to the nucleus, had to be working by driving the expression of an unknown secreted protein able to signal back to the *wingless* cells. Sorting out the specifics of the *wingless–engrailed* reciprocal interaction became the focus of Phil's work when he moved over to Oxford, and he started off by cloning *patched*, another segment polarity gene that his work at the LMB had shown to negatively regulate *wingless*. By running the Patched protein sequence through a database containing information on all currently known proteins, it was clear that it looked like a transmembrane protein, looping itself 12 times through the cell membrane like a particularly elaborate granny knot.

In 1991, Phil's lab published a *Nature* paper on the *wingless–engrailed* loop, which he remains particularly proud of, and in which the subject of this chapter finally makes an appearance. Although *patched* was clearly part of the *wingless–engrailed* story, there had to be more to it. Because *patched* repressed *wingless* and, furthermore, was only found in *wingless* and never in *engrailed* cells, the identity of the signal coming from *engrailed* cells to switch on *wingless* in neighbouring cells was still a mystery. There was a further oddity. When *patched* was lost through mutation, the stripe of cells expressing *wingless* was enlarged, and *engrailed* expression also popped up in the wrong places; as expected, loss of *patched* relieved some kind of repression. However, when Phil did the opposite experiment, overexpressing *patched* so that it was on everywhere all the time, he didn't see the opposite result. Too much Patched should have extinguished expression of Wingless and hence Engrailed, but instead, the two proteins looked completely normal.

It was here that Phil's purely genetic background came in useful, allowing him to make an important conceptual leap regarding what Patched was doing. As he remembers: "That's the great thing, if you're a geneticist, you don't know any cell biology or biochemistry, so for me, Patched was just a membrane protein and therefore it could act as a receptor. The fact that it was a 12-pass transmembrane protein and didn't look like a receptor wasn't a problem for me." Once Phil decided that Patched was likely to be a receptor, it seemed reasonable to propose that its ligand would be the mystery secreted protein coming out of *engrailed* cells. Again, as he was completely free of orthodox biochemical notions that when ligands bound receptors they always switched the receptor on, Phil, going by the genetics, reasoned that when the mystery ligand bound Patched, Patched was switched off. This fitted perfectly with his data—if there was no Patched, there was no Wingless repression, but if there was too much Patched, it wouldn't matter if there was enough of the mystery ligand to keep it switched off, and therefore keep *wingless* on. The regulatory loop as Phil saw it was therefore: *engrailed* switches on a mystery protein; the mystery protein is secreted and goes next door to bind Patched; Patched is switched off; *wingless* comes on; Wingless protein is secreted and makes a return call on the next-door cell to activate *engrailed*.

Phil now knew what he was looking for: a protein, able to be secreted, only made in *engrailed*-expressing cells, which when mutated produced a phenotype resembling loss of either *wingless* or *engrailed*. Fortunately, thanks to Janni and Eric, he didn't have to assemble his own list of suspects; the perpetrator was almost certainly going to be in their list of segment polarity mutants, just like *wingless* and *engrailed* themselves. It was. Figure 2 of Janni and Eric's original *Nature* paper shows three segment polarity mutants, all with defects in bristliness and size. *hedgehog*, the shortest, fattest prickliest one, matched Phil's criteria perfectly.

By the time Phil got interested in it, *hedgehog* had been worked on by a few groups, notably that of Jym Mohler, an ex-postdoc of Eric Wieschaus's who was now working at Barnard College in New York. Mohler had performed an extensive analysis of the *hedgehog* mutant phenotype during fly wing development and had made some crucial observations: *hedgehog* was only on in *engrailed*-expressing cells in the wing and appeared to be a signal, acting not on the cells in which it was made, but upon next-door cells. Furthermore; *hedgehog* mutants looked very similar to *wingless* mutants. A conversation with Eric Wieschaus got Phil even more excited about *hedgehog*; Eric told Phil that embryos in which both *hedgehog* and *patched* were mutated looked just like *patched* mutants, which implied that if the two lay on the same pathway, *hedgehog* was upstream of *patched*, just as might be expected according to Phil's hypothesis.

As proof of his hypothesis, Phil set up crosses to generate fly embryos deficient in *hedgehog* alone or in both *patched* and *hedgehog*, and looked at expression of *wingless*. Sure enough, in *hedgehog* mutants, *wingless* expression faded away completely, but in the double *patched:hedgehog* mutants, *wingless* was on inappropriately, just as in the ordinary *patched* mutant. Although there was still a lot of work to be performed filling in the details, the main players in the regulatory loop had been defined, together with an entirely new type of signalling pathway. Hedgehog protein, made in *engrailed* cells, bound to the Patched receptor and switched it off, resulting in *wingless* transcription.

Phil carried on the *hedgehog* and *patched* project in flies, showing in a single-author *Cell* paper in 1993 that overexpressing *hedgehog* produced a beautiful *patched* phenotype. He also added in a new component of the signalling pathway—the *fused* kinase, which acted downstream from *hedgehog*. However, by that time, he also had other fish to fry, in the most literal sense; the zebrafish, *Danio rerio*, had entered his life in a big way, and the fly biologist's fly biologist had got some backbone into his lab.

It had all started in 1986 during a visit to Janni Nüsslein-Volhard in Tübingen, to discuss some work that Phil was doing on *even-skipped*, the fly pair-rule gene. Janni, much to Phil's surprise, had been rather dismissive of the *even-skipped* data and, instead, had a new enthusiasm that she wanted to talk about. The fly work in her lab had been going amazingly well, with a succession of very impressive postdocs and graduate students studying the many mutants that had come out of the mutagenesis work. Many of these alumni were now running extremely successful labs of their own, based on projects that they'd taken with them from Janni's lab, and as the field she had founded could now do perfectly well without her, Janni was looking for something new and exciting to try. Vertebrate development seemed like an interesting enough challenge, and furthermore, in the zebrafish, Janni thought she had found a vertebrate model organism in which she could repeat the mutagenesis strategy that had been so wildly successful in flies.

Phil soon found himself carried along by Janni's evangelism:

> I got into fish because Janni told me to! She really wanted to talk about fish—she drove me to a pet shop in Stuttgart (never changing out of third

> gear!) to see some zebrafish, as they didn't have any in the lab—it was that sort of level of enthusiasm. I think I felt that we'd answered all the really interesting questions in flies, and the rest of it would just be filling in the details. There was a sense then that because I liked genetics I didn't really want to get into the molecular mechanisms and things, so the idea of having to start doing biochemistry and understanding what proteins did was just not ... So it was more interesting to go into a new system, to be there right at the start, and to start doing genetics again. Janni's charismatic, and you believe it if Janni says something is really exciting. She doesn't do things lightly, so I knew that if she was going to get into fish she was going to create a field that was going to be well supported and dynamic and innovative, and it would not be bad to be part of that.

> I went back and obviously I did read a few papers to see what it was about, and yes, I was taken with fish. I could see that vertebrates presented some interesting developmental problems that we couldn't easily study in flies. At the time I thought *Xenopus* was a really interesting embryological model but it was totally intractable genetically, so it was clear to me that fish were the answer to the genetics bit.

As a fish starter, given that homologues of *engrailed* and *wingless* had been shown to exist in vertebrates (vertebrate *wingless* was shown in 1987 by Roel Nusse's laboratory to be the *int-1* oncogene, and to avoid schizophrenia, ended up being renamed *Wnt1*; in all post-1987 references, *wingless* and *int-1* therefore appear in their new *Wnt1* guise), Phil's lab began to look for zebrafish homologues of the fly genes in which he was interested, keeping a few tanks in the cramped confines of the DBU.

About 2 in. long, silver, with darker stripes down their sides, zebrafish are a classic aquarium staple, both for their prettiness but also because they are as tough as old boots and will survive most forms of benign neglect. They reproduce quickly, with a 12- to 15-wk generation time, and adult females lay hundreds of eggs at a time that develop rapidly and synchronously outside the mother and are transparent, making it very easy to see what is going on during embryogenesis and to identify mutants. By 1980, George Streisinger in Eugene, Oregon had worked out a way of doing genetics in zebrafish, and after his early death from a heart attack whilst scuba diving in 1984, his Oregon colleagues, particularly Chuck Kimmel, carried on developing new genetic methodology. Janni had followed the zebrafish work with interest, and at the close of the 1980s, finally felt that the time was ripe to set up a saturation mutagenesis project. The logistics were even more stupendously daunting than for the fly screen, but as head of a wealthy Max Planck Institute, with *carte blanche* to embark on large-scale projects almost irrespective of cost, Janni was in a perfect position. From the small beginnings that Phil had witnessed, the zebrafish operation in Tübingen ballooned, culminating in the building in 1992 of a massive aquarium for a huge mutagenesis screen.

One of the perks of working at a Max Planck Institute is to be able to use one of the poshest conference venues on the scientific circuit. Schloss Ringberg, built between the wars in a collaboration between the frankly rather peculiar Duke Luitpold of Bavaria and the artist and designer Friedrich Attenhuber, is a grand blend of art nouveau and romanticism, an imposing pile only slightly marred by its sinister history. Duke Luitpold became so obsessively fond of his in-house designer that the only way Attenhuber was able to leave the castle was by committing suicide in 1947, jumping from one of its many towers. Attenhuber's revenge, a series of paintings of the duke executed in the best *blut und boden* Nazi style, in defiance of the Duke's loathing of the Nazi regime, still hang in the castle. Following the Duke's death in 1973, his monumental folly passed to the Max Planck Foundation, and these days, lucky conferees relax in its sumptuous bedrooms, wander the manicured grounds, and play table tennis in its towers.

In spring 1992, the castle was host to about 30 developmental biologists, hand-picked by Janni as the brightest and best in the business. The conference was designed to introduce everyone there to the wonders of the zebrafish and to give Janni a clear idea of what mattered in vertebrate developmental biology, so that she could shape her own future research accordingly. The participants came from all sorts of different backgrounds and ranged from the grandees of classical developmental biology, Lewis Wolpert and 2012 Nobel Prize winner John Gurdon (who both proved to be insanely competitive when it came to table tennis), to young rising stars picked from the fly, frog, fish, chick, and mouse worlds. Phil, as an early adopter of all things fishy, a friend of Janni's, and a big deal in the fly world, was, of course, present. By all accounts, it was a successful meeting; Jim Smith, now Director of the National Institute of Medical Research, but then just hitting his stride at the front of the developmental biology pack, remembers it as "really good natured and academically very rigorous—talks, long discussions and fun. It was held at the right time because developmental biology was so exciting at the time. We drank a lot, ate a lot; it was great."

The Ringberg meeting was the birthplace of a collaboration that transformed the fortunes of the humble *hedgehog* gene and produced one of the most important discoveries in developmental biology. Unfortunately, those involved in this momentous event can't seem to agree on what transpired. What is definitely known is that at Ringberg, Phil Ingham met Andy McMahon and Cliff Tabin, and the three of them got on. How they got talking about *hedgehog* is more open to debate—Phil and Andy aren't even sure that they did, although Cliff has a different recollection:

Here's Andy:

> There was this meeting at Ringberg, which was this castle with these very funny paintings in there that looked like they were idyllic Hitler era paintings of people. I don't know why I got invited but Janni invited a lot of young scientists like me who hadn't done anything and a lot of very serious scientists like John Gurdon. I knew Phil from his *Drosophila* work, but I hadn't really met him properly before. I did try to talk with him at a meeting in Bristol in 1988 but he gave me the Cambridge look down the nose [NB: Phil denies this with some indignation!]. I was in great admiration of the work he had performed but I felt like he was a bit of an untouchable. And then we got talking at this meeting and I was saying, "Well, you're the fly person, I want to talk to you—we know where *wnt1* is expressed in mice, you guys have shown that there's this Engrailed–Hedgehog–Wnt1 interaction, we'd like to look for the Hedgehog component in this." And then Cliff, I don't know how Cliff came in. There were a few too many beers!

Ringberg meeting participants. Phil stands *centre left* in a dark shirt holding a beer. Andy (in glasses) is partially obscured just to Phil's *left*. Cliff is kneeling at the front, wearing a lanyard and a white T-shirt with Everest logo. *(Photograph courtesy of Jim Smith.)*

Cliff has a different memory:

> I'd previously been in contact with Andy but I'd never met Phil—I'd read his papers but never met him. I introduced myself to them at lunch one day and said, "Phil, I'd like to ask you a few questions about your work on Hedgehog because my laboratory is trying to clone some vertebrate homologues." Phil's a great guy, a very solid person, and he said, "I'll tell you whatever you want to know, but I want to be up front with you. You should know that I'm also trying to clone vertebrate Hedgehog." And Andy was sitting right next to Phil and he said to me, "I'm actually cloning vertebrate Hedgehog, too." There was a moment when I wondered whether everyone in the dining room was cloning Hedgehog genes! But it was just a coincidence, it was really just the three of us.

Phil, although lacking in *hedgehog*-specific memories from Ringberg, is very clear on why the Hedgehog Three eventually decided to collaborate; he and Andy started off as the Hedgehog Two but couldn't get any funding:

> This is how Cliff came into the collaboration: Andy and I were particularly excited by work in mice that Laure Bally-Cuif, then a PhD student in France, had just published, which showed that grafts of Wnt1-expressing cells from the midbrain–hindbrain boundary into the forebrain could induce Engrailed expression. This clearly implied a functional conservation of the Wnt1–Engrailed interaction from flies to mice, and on this basis we thought it likely that the reciprocal relationship would also be conserved and be mediated by a vertebrate Hedgehog gene. This was the basis of our proposed collaboration—as it was entirely based on signalling in the brain, there was no obvious reason to involve Cliff. Money was getting tight at the ICRF at the time so I suggested we apply for an HFSP (Human Frontiers Science Program) collaborative grant. Andy flew over to the UK in June or July I think and we spent the day drafting the grant on my dining table in Summertown, Oxford. Because Andy had good secretarial support, he volunteered to complete all the paper work and submit the grant when he got back to the US. One or two days before the deadline, he called me in a panic—he had just read the HFSP fine print that stated that applications would only be considered from citizens of two or more countries. Because Andy still held UK citizenship, it seemed we were ineligible. So Andy said: "You remember Cliff Tabin from Ringberg? He's interested in cloning homologues of fly genes to look for their expression in the limb. We could ask him." I agreed immediately, since Cliff seemed like a smart and fun guy. The rest, as they say, is history!

Andy McMahon, 2013. *(Photograph courtesy of Andy McMahon.)*

So who were these would-be intruders into the Hedgehog field? Andy McMahon, like Phil a Liverpudlian, had been a Zoology undergraduate at Oxford, but had decided after taking the spectacularly good embryology practical course there to become a mammalian embryologist, and had duly done a PhD with Marilyn Monk in London, working on mice. He'd then realised that it "was clear if you wanted to learn about development you had to do recombinant DNA work. I needed to learn how to clone genes and learn all the techniques for molecular biology without losing development. At that time the two systems using molecular biology were *Drosophila* and sea urchins. I'd had an undergraduate project working on fly genetics that I'd found incredibly boring, so I chose sea urchins as an alternative."

Without realising that his chosen laboratory and institution were amongst the best in the world ("I was incredibly fortunate because I fell into lots of things in my life, and I just fell into a great lab at a great place. I don't know why I chose it."), Andy went off to do a postdoc with Eric Davidson at Caltech and learned an awful lot about sea urchins and molecular biology. After a very successful time in the States, he was recruited to the National Institute for Medical Research at Mill Hill in 1984, returning to his first love: mouse embryology. Having got on board an extremely talented right-hand man, Dave Wilkinson, Andy started thinking about what to do.

Inspiration came from *Drosophila*: "We were very influenced by Janni's work. If you looked at the outcome of the *Drosophila* screen, what you could see was that there were genes expressed along the anterior–posterior axis that gave rise to different patterns. You could imagine that if you had taken those embryos at the right time and cut them into bits, and then screened at the RNA level, you would have found those same genes. So we thought well, we can take mouse embryos and do the same." Unfortunately, the level of resolution for cloning did not match Andy's aspirations; the biggest difference between the head and tail ends of a mouse embryo is that blood is made in the tail end, and all the lab ended up cloning was a bunch of globin genes.

Drosophila having let him down, Andy resorted to a much more reliable muse: his wife, Jill. Jill was working downtown at the ICRF on a singularly boring transcription factor called Myb, but in the next-door laboratory, Gordon Peters and Clive Dickson were studying two other oncogenes, *int-1* and *int-2*. At that time, the homology between *int-1* and *wingless* was yet to be discovered, but the genes were intriguing to a developmental biologist:

> We're a multicellular organism, so the interesting genes are ones that are involved in sending information from one cell to another. What would be the other characteristics of interesting signalling factors then? Well, those signalling factors would have important regulatory properties in other contexts. What other contexts? Well, cancer. Int-1 and Int-2 looked like secreted factors. Nobody had found any expression for them in adult tissues, so they were complete orphans, and I thought well, nobody's looked in the embryo. Dave did in situs in the embryo and then we saw these fantastic patterns of expression.

Andy's hunch paid off, and he and Dave, together with Juliet Bailes, published a *Cell* paper in 1987 that showed that *int-1* was expressed in particular cells in the central nervous system during mouse development. They shared the discovery with Harold Varmus and his postdoc Greg Shackleford, whose paper also appeared in the same issue of *Cell*, which has Dave's beautiful in situ photo on the front cover.

Having got into development of the nervous system via *int-1*, Andy stayed there. His next experiment was probably his most spectacular to date. While he was at Caltech, he had become good friends with another postdoc there, Randy Moon, who was now a frog person working in Seattle. Thanks to Roel Nusse's paper, it was now known that *int-1* and *wingless* were homologues, and thus, because *wingless* clearly had a profound effect on fly development and *int-1*'s pattern of expression was intriguing, it seemed like a good idea to see what *int-1* might do to development of a vertebrate embryo. Andy took a series of *int-1*-expressing constructs over to Seattle, and Randy injected them into frog eggs. The result was completely unexpected and very exciting, although not everyone was convinced. Andy remembers he "went to almost all the amphibian people and they said 'Oh God, it's some bloody artefact.' The only person who was the least bit encouraging was John Gerhart at Berkeley, and he said, 'No, this looks interesting.' " It was. What Int-1 was doing was causing an axis duplication in the frog embryo—instead of one tadpole with one backbone forming, *int-1*-injected embryos were trying to make a two-headed tadpole. Int-1, a single protein, was able to mirror the effects of one of the most important groups of cells in an early amphibian embryo: Spemann's Organizer.

The study of vertebrate development was way behind fly embryology molecularly, but in terms of general concepts, vertebrate embryologists had a pretty good idea of what was going on. Just like flies, all vertebrates start off as an egg and from there must specify a head-to-tail (anterior–posterior) axis and a back-to-belly (dorsal–ventral) axis. The early vertebrate embryo takes shape in an exquisite ballet, in which the cells of the primitive streak create a long furrow in the spherical epiblast through which other cells pour inward, laying down an anterior–posterior axis to eventually form the notochord, over which the neural tube, the precursor of the spinal cord and brain, is created. In 1924, Hans Spemann and Hilde Mangold showed that if cells were taken from one particular region of an early newt embryo and grafted into another newt embryo at the same developmental stage, those cells caused the host embryo to attempt to develop two heads and two spinal cords; in other words, to duplicate its anterior–posterior axis. The region of the embryo was named Spemann's Organizer. (Poor Hilde Mangold died of burns in 1926 after being caught in a gas explosion whilst heating milk for her baby son, and it is a great shame her codiscovery does not bear her name too.) Spemann's Organizer was the first example of an inductive interaction in development, where a signal from one group of cells influences the development of an adjacent group. Analogous organizers are found in all vertebrates, not just for anterior–posterior axis specification, but for many other processes, such as limb formation.

Organizers require messengers to pass their instructions on to neighbouring cells, and the vertebrate embryologists dubbed these messengers *morphogens*, defining them as secreted molecules that can induce changes in cell fate. Morphogens were hypothesised to move by diffusion from the organizer, and their effects would be concentration dependent; high concentrations of morphogen, found near the organizer cells, would dictate different cellular outcomes from the lower concentrations found further away from the morphogen source.

Returning to Andy and Randy's improbable result, it was not surprising that the frog people were sceptical because the data suggested that Int-1 might be one of the fabled morphogens, or at the very least, part of the patterning systems involved in axis specification. The discovery was a landmark in developmental biology, the first name on a hitherto blank molecular page, and McMahon and Moon's 1989 *Cell* paper created quite a stir. Leaving Randy to his frogs, Andy returned to London intent on really nailing what Int-1, now morphed into Wnt1, was doing in the mouse central nervous system. Once Phil published his 1991 paper showing that Hedgehog and Wnt-1 were inextricably connected during fly development, it was just a matter of time before Andy realised that he needed to get some Hedgehog into his life.

In contrast to Andy and Phil, bred in the developmental biology laboratories of the United Kingdom, Cliff Tabin was a very different animal; a Midwesterner, he was a molecular biology star before he had finished his PhD, when amongst other things he discovered, whilst in Bob Weinberg's lab at MIT, that a normal cellular gene, *Ras*, could be converted into an oncogene by making a single mutation in its DNA sequence. Blessed with exceptionally green fingers at the bench, where his knack for making impossibly hard cloning experiments work was a Boston legend, he had moved over into vertebrate development for a postdoc, initially working in Doug Melton's lab at Harvard on homeobox genes in frogs. However, his experiences as a graduate student made the Melton lab's experimental approach rather unappetising:

> Doug's idea (and other people's) was to clone a bunch of these genes, inject them into frog eggs and look for a phenotype that you could interpret. But I was in Bob Weinberg's lab when they were banging their heads

Cliff Tabin and Doug Melton, 1997.
(Photograph courtesy of Cold Spring Harbor Laboratory Archives.)

> against the wall trying to figure out what Ras did. Once you have a protein, in those days you couldn't knock it out, and biochemistry was hard. Maybe you get lucky, but it made more sense to me to turn that around. Instead of saying, "Here's a protein, what does it act in?" you say, "Here's a problem, what acts in it?"

Given that the homeobox genes on which he was working were involved in the specification of segment identity in flies, Cliff thought it would be best to start with something similar: "What is segmented about us? Two things, the vertebrae and the limbs. There is no classic somite literature, but there is a vast limb literature. So I said to Doug, 'Why don't we try and look at limbs and see if there are any homeobox genes involved in limb development?' And he thought it was a good idea, but you couldn't do it in frogs as they go weeks before they get limbs. He said, 'If you want to do that you should do some limb regeneration in newts, but I don't want to do that in my lab.' "

Doug recommended that Cliff applied for an independent position that had just come up in the Massachusetts General Hospital Department of Molecular Biology, throwing him in at the deep end, but meaning that if anything came of his work, he alone would get the credit for it. The strategy was a bit risky, but Cliff didn't really have any alternative; nobody in the world was doing what he was proposing.

Limb development had been studied in two ways: by chopping the legs off newts and watching how they regenerated, and by using chick embryos. Starting off with the former, Cliff went to learn some basic techniques with Jeremy Brockes in London, swapping his molecular biology know-how for tips on how to amputate legs so that they stood the best chance of regrowing. London was a good place to be for a newcomer to the limb field, because it was not only full of three-legged newts in various stages of recovery, but also harboured Lewis Wolpert, whose work on chick limbs had shaped current theories regarding how organizers were able to govern development. Lewis, a brilliant theoretician, had been less surprised than most at the discovery of the many homologous fly and vertebrate genes: "I decided that if evolution had taken the trouble to have general principles for the genetic code there were absolutely going to be general principles for development. So I wasn't looking for specific solutions, I was quite convinced that there was going to be some general answer" (Smith 2000).

The Wolpert fondness for the general rather than the particular had not stopped his lab from publishing some of the most important work on limb development, and happily, he and his people were very friendly to the molecular biology interloper, making up for some of the American limb supremos, who'd been less than kind to Cliff: "I had my first serious discussions with Lewis at a conference in Santander shortly after I'd been in London, and he welcomed me into the field, albeit with the same paternalism as he treated others working on the limb: 'Very interesting experiment, my boy, but you really don't understand it—let me explain your data to you!' "

Lewis's lab had been busting a gut for many years trying to determine exactly how the organizer of anterior–posterior patterning in the limb did its job. Limbs have two main organising regions, found in the limb buds, which are little nubs arising from the main body of the vertebrate embryo. The apical ectodermal ridge is responsible for getting the limb to the right length with the correct series of bones in it. On the posterior side of the limb bud, the zone of polarising activity (ZPA) determines the anterior–posterior axis, controlling the number and order of fingers and toes. In 1968, John Saunders and Mary Gasseling showed that grafting a ZPA from one chick embryo onto the anterior side of the limb bud of another chick embryo resulted in a limb with double the number of digits (six instead of three; chickens lack digits 1 and 5). The extra three digits were formed as an axis duplication, a mirror image of the normal three; the usual pattern of 4, 3, 2 was changed in the grafted embryos into 4, 3, 2, 2, 3, 4. The following year, Lewis proposed that, as for Spemann's Organizer, a morphogen diffusing from the ZPA was responsible; the morphogen would be present at progressively lower concentrations as it diffused further from the ZPA. High levels would specify digit 4, intermediate levels digit 3, and low levels digit 2.

In 1975, Lewis's two colleagues Cheryl Tickle and Dennis Summerbell provided data consistent with Lewis's model; by shifting where they grafted the foreign ZPA, they could induce different numbers and types of digits, but always in order, and always with digit 4 forming nearest the ZPA, and digit 2 furthest away. And there things stuck for some time. Nobody had any clue what the morphogen might be, and so trying to purify it biochemically was well-nigh impossible. Plundering the Wolpert lab's stocks of random chemicals to see whether they worked as morphogens led to the discovery in 1982 that retinoic acid and related compounds could cause the axis duplication, but by the time Cliff arrived on the scene, there was some evidence that retinoids might not be the real deal; instead of being a product of the ZPA, as the ZPA morphogen had to be, they seemed instead to be instructing the embryo to make an entirely new ZPA.

Having imbibed newt lore from Jeremy Brockes and limb theory from Lewis and his lab, Cliff started well in the limb world, cloning a slew of vertebrate homeobox genes and showing that some were expressed in interesting places during newt limb regeneration. However, after this early success, things got a lot harder. To figure out which of the genes might really matter, the obvious next step to take was to overexpress them during limb regeneration to see whether the process could be perturbed. The problem was that working out how to do this was not trivial, even to a molecular biology jock of Cliff's undoubted calibre. In his other claim to PhD fame, Cliff had been one of the first people to develop and use retroviral vectors. Derived from RNA tumour viruses, the vectors carried foreign genes into mammalian cells, depositing them in the host genome, but they were dependent on particular receptors to allow them to enter cells. Cliff wanted to use such vectors to infect newt limb buds during regeneration, but newts, inconveniently, didn't have the right receptors. After some frustrating attempts to make newt-specific vectors, Cliff embraced with relief the development of chicken-specific retroviral vectors by one of his old oncogene stable mates, Steve Hughes. Abandoning his newt tanks for egg incubators, Cliff embarked on a new life as a chicken person at Harvard, who had by now decided that he was good enough to be given a permanent job.

Good developmental biologists are remarkably adaptive beings. Immersed in the language and customs associated with their own pet organism, they are also happily multilingual, able to converse in Fly, Fish, Frog, Mouse, or Chicken and sundry minority dialects according to circumstance. Part of this is due to the commonality of development, that at early points there are so many parallels, but part is a very special skill, picked up almost by osmosis, relying on a superb visual memory and an ability to screen out unimportant differences. However, before the imperative of molecular biology unified the fields, there was both a temperamental and an experimental divide between the fly biologists, with their rigorous genetic training and tranquil, ether-scented laboratories, and the rest. For a fly person of Phil Ingham's calibre and devotion, making the jump from flies to vertebrates had therefore been quite traumatic, and his transformation was certainly not complete by the time of the Ringberg meeting:

> It was horrible! I did a little bit—I helped my postdoc Uwe Strähle set up the aquarium and I looked after all the fish while he was on paternity leave, but I couldn't wait for him to come back! It was just completely different. Whereas with flies you spend 80%–90% of your time thinking or doing experiments and 10% looking after them, with fish it was the other way round. Cleaning out tanks, feeding them ... we weren't actually *doing* anything. And of course I totally underestimated what was required if you wanted to do genetics in fish, and that I was never going to have a hope of doing mutant screens. Just the infrastructure that you need—you need a huge fish facility and an army of people. The idea that I could do it was just ludicrous, but I didn't see that. So of course what we ended up doing was basically what the frog people were doing, just cloning homologues and looking at their expression. Mostly what I did was just transferring in situ protocols from flies to fish. I was still doing nearly all fly stuff myself.

Phil clearly needed something to galvanize his flagging passion for fish, and, despite being separated from them by the Atlantic (Andy had by this time moved on from NIMR to a faculty position at the Roche Institute in New Jersey; from there he went to Harvard in 1993), having two enthusiastic new vertebrate collaborators was a welcome stimulus. Furthermore, Andy and Cliff were sufficiently different from Phil and each other, in both style and outlook, to make the prospect of a collaboration rather interesting. Phil hid his considerable intellect beneath rumpled linen outfits and an amiability of demeanour more often to be found punting on the Cherwell on a languid summer afternoon than in a science lab. In contrast, Andy, a fanatical long-distance runner with the physique and dogged outlook to match, was a ruthlessly analytical scientist whose sometimes alarming seriousness was lightened by a surprisingly sweet smile and a tendency to unexpected laughter. Cliff, summed up to his great embarrassment in *Natural Obsessions*, Natalie Angier's book about the Weinberg lab, as "delightfully handsome in a husky, masculine, linebacker sort of way," was different again. With a seemingly endless capacity for absorbing new information, he was astute, intellectually generous, and addicted to the adrenaline hit of getting a really good result: "By personality there are people who don't want to get too excited, they don't want to get ahead of themselves, and by the time they've verified enough that they believe something, it's so anticlimactic they never get excited at all. I'm much happier saying 'that's great!' and getting *really* excited and then if it's not great, all right, then I'll move on, and not worry about it. The lows in that aren't too bad."

The plan hatched by the three collaborators was a simple one: determine whether there were fish, mouse, and chick versions of the *Drosophila hedgehog* gene, find out where and when during development the vertebrate *hedgehog* was expressed, and do some kind of functional assay to figure out whether it was important or not.

From the first, there was considerable incentive to work quickly. As Cliff says: "We knew we had an opportunity and it was exciting. We had the sense that it wasn't that brilliant an insight to try and look in vertebrates for *Drosophila* genes and that sooner or later, other people were going to do it." Fortunately, the three collaborators got a head start on the competition, as Jym Mohler, having completed the detailed study of the fly *hedgehog* phenotype that had previously been so useful to Phil, had moved on to trying to clone the gene itself. Mohler was a very nice guy, had been a classmate of Cliff's at MIT, and was friendly with Andy and Phil. He was more than happy to let them have what fragments of the fly *hedgehog* sequence he had managed to clone to date, even though, as Andy puts it, there were just "weird bits and pieces."

The McMahon lab set to work with the chunks of the fly *hedgehog* gene, using them as probes to try to fish out similar sequences from a genomic library constructed from mouse DNA. Quite quickly, they hit pay dirt; the fly *hedgehog* probe lit up one of the genomic clones, and it was clear that the collaboration was going to have something to work with. Andy posted his new clone off to Cliff and Phil so they could take a look in their respective organisms for something similar, and in the meantime, started doing in situs to find out where during development vertebrate *hedgehog* was expressed. His data came as something of a disappointment. Rather than being switched on in the brain, as Andy and Phil had hoped, the gene was coming on at its highest levels in a place where, even with the best will in the world, it would be difficult to claim there was much thinking going on—adult male testes.

If the location of the Hedgehog homologue was less than ideal for Andy and Phil, it was much worse for Cliff. In contrast to chicken eggs, in which chick limb experiments are conducted, fully grown boy chickens are rather hard to come by: "It was so deflating—it wasn't going to be the brain, it wasn't going to be the limb, and it was especially bad for us, because for us, we've got to use roosters. We got a couple, just to show it was in the testes, we had to go to a butcher's shop and get these live roosters. We sacrificed them—it was a disaster." Out of the seeds of desperation, Cliff had a very good idea: "Because it was not what we were hoping for, I told my postdoc Bob Riddle that maybe there really is only one, but it *could* be a gene family, it could be more than one. So Bob looked and found that there were three Hedgehog genes, a family of three. As soon as I knew, we let the other guys know."

Having found three Hedgehogs, the question arose in the Tabin lab of how to differentiate between them. Finding new genes is a great excuse for exercising a bit of creativity in naming them, and Cliff and the lab rose to the challenge admirably:

> I gave them provisional names for fun. I started off naming the genes for species of hedgehogs found in different parts of the world, but the only one we still use is *desert hedgehog*, which was the original one Andy found in testis. We had to change *short-eared hedgehog* to *Indian hedgehog*, as short-eared sounded like a phenotype. I was originally calling the third gene *common european hedgehog* as it was in the most places and was common! But Bob Riddle had a friend in England where they were marketing this video game called "Sonic the Hedgehog." Bob's daughter Annie had her sixth birthday and the friend sent her a "Sonic the Hedgehog" comic book. The other thing about Bob was, he played bass guitar in a rock band when he wasn't at work, so the idea of calling a gene *sonic hedgehog* really appealed to him.

Somewhat to Andy and Phil's British dismay, *desert* and *indian hedgehogs* were duly joined by *sonic*, but both comforted themselves with the thought that the naming was as yet only provisional.

In the midst of the christening ceremonies in the Tabin lab, Phil and his lab had been busy with Andy's original clone of *desert hedgehog*:

> Andy sent the clone to us, and Stefan Krauss, my postdoc, pulled out what actually turned out to be a full length copy of fish *sonic hedgehog*. While we were doing that, Andy's postdoc did the in situs, and saw no expression in the embryo, but expression in the adult testis. So Stefan just put the clone he'd isolated on one side because he thought it wasn't going to be of any interest. Then Cliff isolated three different bands. He immediately told us that they'd got three different bands, meaning three different genes. So then Stefan immediately went and got his clone, made the probe

and did the in situ and that's when we found this amazing expression pattern.

Stefan's amazing in situs unexpectedly became the centrepiece of a social, rather than a scientific event in the Ingram lab:

The day that he did the in situs I had a visit from fundraisers. So I had these four or five little old ladies in the lab and I showed them round, and Stefan came in with the dish of embryos. So I said, "Why don't we look at these fish embryos?" So a fundraiser was actually the first person to see the *sonic hedgehog* expression pattern—she looked at it before I did. I looked, and went "Wow!" But the only other zebrafish gene that we'd cloned then was *axial*, and *axial* has a pattern of expression very similar to *sonic*. So when I looked at it, I thought Stefan had mixed up his probes. Of course I didn't say anything in front of those ladies, but as soon as they'd gone I said "You must have mixed up your probes Stefan, this is *axial*." But then we looked more carefully and realised there were subtle differences. So then Stefan just did a drawing of the expression pattern and I faxed it to Cliff and Andy.

The *sonic hedgehog* expression pattern was, quite simply, a game changer for developmental biology. In the developing zebrafish embryos, *sonic* was first seen in the inner cell layer of the embryonic shield, the region corresponding to Spemann's Organizer in amphibians. As the embryos grew, *sonic* then appeared in the notochord, the structure over which the neural tube was laid down, and then came on in the floor plate, a thin strip of specialized cells overlaying the notochord and running like a seam along the ventral side of the nascent spinal cord. All these three places—the node, the notochord, and the floor plate—shared one characteristic: they were all fundamentally important organizers, all able to dictate the fate of neighbouring cells by secreting morphogens, the mysterious messengers that nobody had been able to identify. And in each of these organizers was *sonic hedgehog*, a close relative of a secreted molecule known in flies to be able to change the fate of neighbouring cells. It looked tantalisingly likely that *sonic* and the morphogens were one and the same; in a remarkable testimony to recycling good ideas, evolution was using the same molecule in multiple different places.

Cliff, the limb man, was as excited as his more neurally inclined collaborators:

As soon as I saw the floor plate and the notochord expression I knew we had the ZPA factor. The people in my lab were laughing at me but I was absolutely convinced. Other people had already shown that bits of the

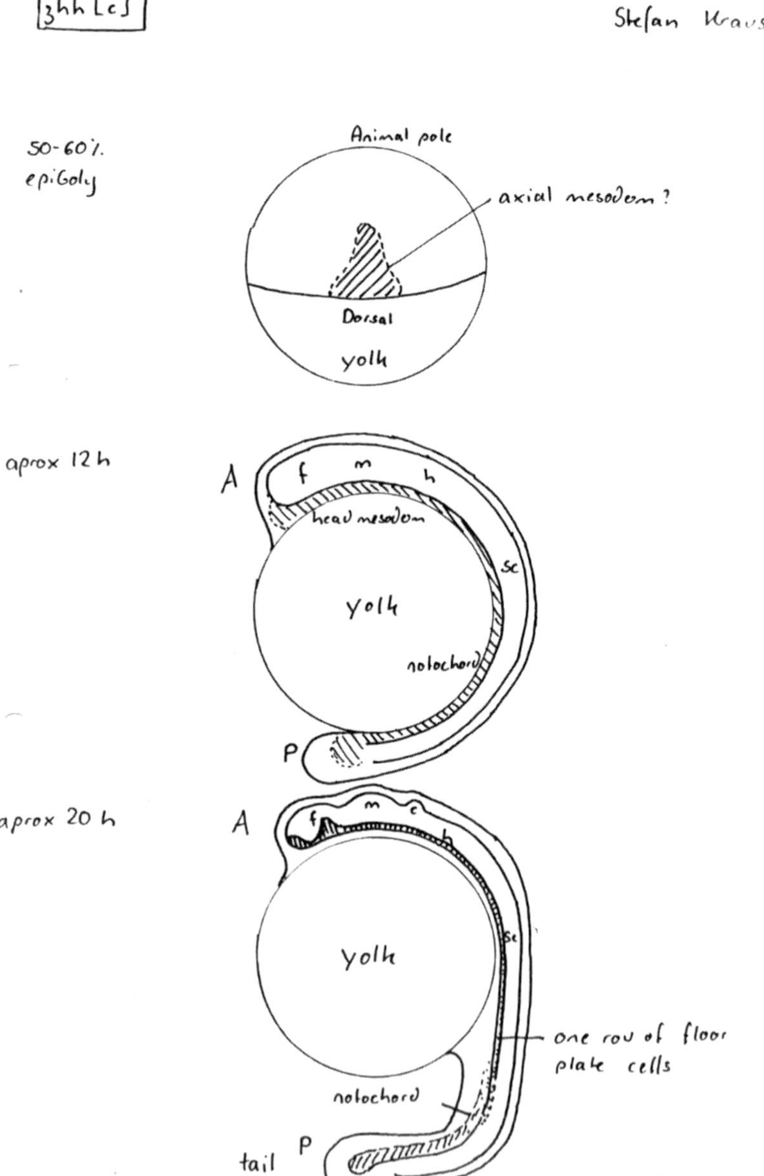

Stefan Krauss's original sketch of *sonic* expression in zebrafish, faxed by Phil to Cliff and Andy. *(Photograph courtesy of Phil Ingham.)*

floor plate and the notochord could cause ZPA-like duplications in the limb, that there was a conserved signal. I thought it just had to be right. We knew that *sonic* was expressed in the limb bud—we'd isolated *sonic* RNA from limb buds, so I didn't even need to see the expression pattern. My postdocs thought once they saw the expression pattern, that OK, *sonic* was part of the story. But I said "No—it *IS* the story!—what are you talking about, *part* of the story?!"

In retrospect, Cliff is anxious to emphasise that his conclusions were actually very naïve: "There are many genes expressed in the notochord and neural tube, and several of them are also expressed in interesting and important ways in the limb. I jumped to a conclusion because that's my nature. In this case, of course, I happened to be right. But that was lucky, not clever intuition."

Whatever Cliff's opinion of his younger self, further in situs from all three organisms soon confirmed Phil and Stefan's data and proved Cliff to be correct: *sonic* was found in the node, notochord, floor plate, and limb (or fin) ZPA in fish, mouse, and chick embryos. Andy, not given to hyperbole, summed up his feelings: "It was one of those times that are probably unique in people's scientific lives where you do an in situ hybridisation and you instantly see the two most interesting organizing areas of the embryo. All the classical literature said the limb, the ZPA, very important, and the notochord and floorplate, very important areas. It was pretty astonishing, I have to say—really a gobsmacking moment. You have to have at least one of those in your life!"

Jim Smith, who had spent his PhD in Lewis Wolpert's lab in the late 1970s laboriously performing ZPA grafts in an attempt to learn more about the morphogen, remembers seeing Cliff at a meeting shortly after the first chick limb in situs were performed: "It was amazing—the most exciting thing I'd ever seen. It confirmed in such a direct way what we'd inferred from all our experiments. The actual seeing of those in situs was it—you'd have to be blind not to realise. When I saw it, if you had given me a photo of a chick limb bud and asked me to colour in the bits of the chick limb that expressed the magic morphogen, it would have been where *sonic* was expressed. It was the most extraordinary thing. I felt very envious!"

By July 1993, the bulk of the in situ studies had been completed. Although it was obvious to all concerned that *sonic* was the real deal, proof was required to really nail it, in the form of overexpression experiments; if *sonic* was really the morphogen, making too much of it in the wrong place should result in the induction of inappropriate structures. This meant that Cliff had to show that *sonic* by itself could act like a ZPA graft in the chick limb, and that Andy and Phil had to use *sonic* to mimic the actions of the notochord and floor plate, which normally induce development of the ventral half of the spinal cord. Everyone went into overdrive, although Andy had a little trouble convincing some of his lab members that nights out were no longer going to be a feature of their lives for a while: "The people in the lab at the time were not necessarily doing things at the right pace and I said to them that this was important, and that they really needed to move on this." Word has it that at the time, Andy may have expressed himself a tiny bit more forcefully.

Although the experiments being done in the three labs were all designed to converge on the same scientific objective, they required different organism-specific techniques, all technically difficult and in some cases novel. Phil's lab, somewhat to his shame ("doing overexpression experiments is a bit embarrassing for a fly geneticist"), was making synthetic mRNA transcripts of the fish *sonic* gene and microinjecting them into fertilised eggs, which were then able to translate the mRNA into *sonic* protein. As long as the tank conditions were favourable, an art in itself, persuading pairs of fish to make eggs was pretty easy, as was harvesting them; a strip of gauze could be left in the tank and simply lifted out together with the eggs that had fallen onto it. Thereafter, things got fiddlier. Hundreds of the tiny eggs were examined one by one under a microscope to determine what stage they were at, then quickly microinjected with the synthetic mRNA, then got back into water, in order to maximise survival. Because nobody had tried to overexpress anything in zebrafish before, Stefan and his fellow postdoc Jean-Paul Concordet were pretty much making it up as they went along.

Andy's lab was also using eggs, but had to do full-on microsurgery to harvest them from pregnant mice and reimplant them in recipient mothers. The eggs, much smaller than those of the zebrafish, were microinjected with plasmid vectors in which expression of the *sonic* gene was governed by the regulatory elements normally driving expression of *Wnt-1* in the midbrain and dorsal spinal cord; such transgenic constructs would switch *sonic* on in completely the wrong place during development of the central nervous system. The technique of making transgenic mice by microinjecting eggs with DNA had been introduced by Ralph Brinster and Richard Palmiter in the early 1980s, but was still a bit hit-and-miss even after 10 years. One had to have a steady hand and a great deal of patience and stamina, and even then, the microinjected eggs were often too damaged to develop, and if they did, the transgene could be silent, or defective.

Cliff's lab, using chicken eggs, had to use a unique blend of very old and extremely new technology. In a procedure familiar to generations of chicken farmers, eggs were first candled, held up to a bright light, to check the position of the embryo. Using staging charts developed in 1952, the age of the embryo was then determined by looking at it through a small 1-cm^2 hole cut in the shell. All grafting to the limb buds was done through this same window, and not surprisingly, was so technically demanding that some people never got the hang of it. The modern twist used by Cliff's lab was that instead of using pieces of tissue from another embryo, their grafts comprised pellets of cells infected by a chicken retroviral vector carrying the *sonic* gene. This made the graft even harder because the pellets had a propensity to fall apart during the implantation procedure. The lab at least smelt nice; to look at the embryos at the end of the experiment, the tissues were cleared of colour using methyl salicylate, the active ingredient of liniment, giving Tabin lab members the air of a particularly injury-prone sports team.

By dint of working their socks off, by October 1993, all three labs had produced unequivocal evidence that *sonic* was exactly what they had thought it to be. Cliff's data were probably the easiest to interpret because the readout for his assay was simple; do embryos get extra digits? They did. The implanted *sonic*-expressing cells in the limb bud could cause axis duplications as if they were a whole ZPA graft, and furthermore, *sonic* could induce expression of the downstream homeobox genes that were known to be involved in subsequent patterning of the limb. Andy and Phil, working in the nervous system, also came up with the goods. In mouse embryos, turning *sonic* on in the dorsal neural tube resulted in extreme abnormalities of development, with ventral markers being switched on. *sonic* was attempting to force the dorsal spinal cord to become ventral, just as expected for a morphogen dictating ventral development. In fish, as in mice, the same illicit activation of ventral markers occurred in embryos overexpressing *sonic*, and there were also forebrain abnormalities. As the icing on the cake, Phil's lab also did a very cute experiment in flies, using fish *sonic* to rescue flies whose own *hedgehog* gene was mutated. This conservation of function across some 700–900 million years of evolution was pretty amazing in itself, but it also suggested that the *hedgehog–patched* signalling pathway that Phil had already worked out in flies was likely to be conserved in vertebrates, providing an incredibly useful starting point for future vertebrate studies.

Having accumulated enough confirmatory data to sink a battleship, the three labs met in Boston to thrash out the details of how their papers should be written up. To fulfil their HFSP grant commitments, the original intention of the three lab heads had been to publish everything they did jointly, with everybody's name on all papers, but the *sonic* results were so mind-blowing that a postdoc rebellion took place in the Tabin lab. As Phil remembers, "When Cliff's lab got their result, Cliff's postdoc said, 'There's no way you and Andy are going to be on a ZPA paper!' "

The Boston discussion on how to carve up the results was quite difficult, with all of the postdocs arguing about it in one place, and Cliff, Andy, and Phil holed up in Andy's office, negotiating in a more civilised manner. Eventually, it was decided to split the data by lab, although Cliff had so many results that he decided to submit two papers, rather than one. A preliminary enquiry from Cliff to *Cell*'s editor, Ben Lewin, indicated that the journal would be more than happy to publish the combined work, as long as the data were solid. The intention was to submit all four papers to *Cell* simultaneously, and soon. Cliff recalls that "We did want to get it out quickly—we were conscious of the end of the year, we were aiming for the December issue, because we felt that over another full year's time, someone else would have got it."

Cliff headed home to Chicago in order to write uninterruptedly: "I was by nature interested in sorting through the classical literature and laying out agendas and so on. I certainly had in the back of my head the information about how to write a paper on the discovery of the signal that was the ZPA. And then the logic of it was very straightforward. It wasn't like we had a lot of confusing data, the data all fitted together."

Back in Oxford, Phil, writing his first ever paper regarding vertebrate development, was not quite so calm:

> I think to be honest, we went from being euphoric and excited about it to then being under intense pressure—it was just something we had to get out. We had to get it to Ben Lewin before some deadline or other. David Ish-Horowicz had a postdoc who was flying to Boston, and so he was going to take it over for us. We were in the Zoology department all night finishing it. And then a fire alarm went off about one in the morning and we had

to evacuate the building. The fire brigade seemed really amazed that there were people in there working at that time! So it was just a bit stressful actually.

The four papers were submitted together in mid-November 1993 and were enthusiastically received by the anonymous reviewers, although Cliff, to his disappointment, was told that he had to condense his two papers into one, because there were not quite enough results to justify the second. The haste with which the papers had been written turned out to be justified. News had leaked out onto the scientific grapevine regarding the results, and other laboratories, notably those of Phil Beachy in Baltimore and Tom Jessell in New York, turned out to be not far behind. Jessell, once he heard about the data, is rumoured to have locked his lab in and told them all that if they worked hard enough, they could be joint first authors on any resulting paper. However, even the Jessell lab on 24-h shiftwork was not quite fast enough to scoop the three collaborating labs, and Phil, Andy, and Cliff's papers were published to great acclaim in *Cell* on the last day of 1993.

In the 20 years since *sonic* was unmasked as the master morphogen of patterning in the ventral neural tube and the limb, much has happened. The signalling pathway in which Sonic and its Hedgehog family relatives sit turned out to be one of a select few universal pathways, proven to be useful early on in evolution and then used by multiple cells and organisms. Hedgehog signalling pops up everywhere that cells need to make decisions regarding their fate based on who their neighbours are. Readouts are dictated by the milieu in which the pathway works; it is more an engine that can be harnessed to drive many different processes than a process in itself. The pathway also bears the added distinction of being one of only two discovered through genetics, rather than by analysis of transformed cell lines, perhaps because the way in which Hedgehog proteins are processed and signal is unique and not easily discoverable by conventional means.

At the molecular level, the Hedgehog signalling pathway is still growing in complexity, although the basics are now mostly understood, from the way in which Hedgehog proteins are cleaved and secreted from their originator cells, to how they bind and repress the Patched receptor, activating the Patched target protein Smoothened to trigger a signalling cascade that regulates transcription of Hedgehog target genes in the nucleus. Recently, it has been shown in vertebrates that the incoming Hedgehog signal is directed into the primary cilium, a sensory organelle that sticks out like an aerial from the surface of virtually all cells in the body, acting as a fundamentally important receiver for multiple extracellular stimuli.

In the wider sphere of development, there are still many prickly Hedgehog puzzles left to contemplate, as the myriad processes in which Hedgehog is now known to act are slowly deconvoluted. Even after 20 years, nobody quite knows what is going on in neural tube patterning or limb morphogenesis or, indeed, how Hedgehog can act as both a long- and short-range morphogenic signal, although for all of these, there are credible, if probably oversimplified, models.

In Jim Smith's view, *sonic* was the Last Great Discovery in the current era of developmental biology:

> These days, there's a lot of crossing T's and dotting I's. There were so many big things around that time in developmental biology—there was mesoderm induction, then the whole homeobox gene stuff in vertebrates and the fact it all matched with *Drosophila* was mindblowing, and then this stuff, which for me solved this big problem. I thought, "What could top this?" It was hard to imagine how the field could go any higher. It turned developmental biology into cell and molecular biology—the fact that you had factors. Nowadays, the big questions are so poorly defined that you can't really answer them—it won't be for another decade before it is even possible to approach them—we'll be into a new era when we can.

Phil, to some extent, agrees: "Well, I'd certainly say that it was the last big thing for my career! I'm not sure it's true of the whole field! It is true in terms of the signals—Hedgehog turned out to be the most spectacular of the pathways. This was the last major untapped signalling pathway that does have a spectacular effect."

What Phil and Jim and, indeed, Andy, definitely agree on is the thorny issue of whether *sonic* was a daft name for a gene. Andy and Phil would have been a lot happier with naming the vertebrate Hedgehog family vhh1, 2, and 3, or some version thereof, and Jim is on record on Radio 4 at the time, tutting about calling something after a video game. However, Cliff is unrepentant: "I like names that are memorable. We were playing off the *Drosophila* name in homage to the original naming system." Phil is still indignant, but a little rueful:

> *Sonic hedgehog* was the first silly gene name. The earlier fly names were quite sensible—they were always descriptive in some way. *Sonic* broke the mould as it was a reference to something completely different. Our paper in *Cell* doesn't even have "Hedgehog" in the title. I certainly didn't use *sonic* as I thought that name would just disappear. And then Andy and Cliff both used it and of course that was a big mistake for us. I think part of the reason our paper doesn't get cited so often is because it doesn't mention Sonic hedgehog.

There's one remaining story regarding the naming of *sonic*, which Cliff told with great relish in an interview in 2008: "The funniest part involved a colleague whose husband was driving home from work. We submitted the papers in October and they were published in the last issue of December; in November Sega started promoting Sonic the Hedgehog in the US, including at McDonald's. The husband saw the big sign, slammed on the brakes and went to a pay phone to call his wife—'You're not going to believe this, McDonald's is doing a promotion on Cliff's gene!'" (Tabin 2008).

In 1996, the concerns about calling a gene with obvious importance in development after a cartoon hedgehog were to some extent justified, when Patched, the Sonic hedgehog receptor, was shown to be a human tumour-suppressor gene. Mutations causing loss of function of Patched were associated with an unpleasant disorder called Gorlin syndrome, and not surprisingly, the comic origin of the pathway's name was rather lost on Gorlin patients and their relatives.

Gorlin syndrome sufferers have multiple developmental defects including very characteristic facial changes: many have wide-set eyes, a "saddle" nose, cleft palate, and pronounced brow bones. As well as having profound prenatal effects, the absence of Patched is disastrous postnatally. Normally, the Hedgehog pathway is only active post-birth in a few very specialised places, such as in renewal of the lining of the gut. With Patched missing, the pathway has no negative regulator, and thus is switched on inappropriately in multiple places. In cells where Hedgehog signalling is the primary driver of proliferation, there is therefore a predisposition to cancerous growth, and Gorlin patients can develop rhabdomyosarcomas, tumours of the muscle, medulloblastomas, brain tumours, and most commonly, a glut of basal cell carcinomas, skin cancer.

Knowing that defects in the Hedgehog pathway caused cancer prompted a mass survey of many different cancer types. In very short order, it became obvious that mutations in Patched and its downstream target Smoothened cropped up in multiple sporadic cancers of the colon, pancreas, and ovary, but were particularly common in basal cell carcinoma, where a majority of cases had either lost Patched activity or had suffered a Smoothened mutation that made the protein less responsive to Patched repression. Caught early, basal cell carcinoma, the most common form of skin cancer, is eminently curable, but if it is allowed to metastasise, it is a killer. In the 1990s, there was no treatment beyond palliative care, and, therefore, trying to find a suitable drug able to switch off the Hedgehog signalling pathway, and perhaps save or prolong lives, was a very exciting prospect.

The possibility that the Hedgehog pathway might be important in disease had not been lost on Phil, Cliff, and Andy, but the idea that it might be good to exploit their expertise and harness it to drug development came from an unexpected source, Cliff's dad. Julius Tabin was originally a physicist at MIT, and in World War II had worked on the Manhattan Project with Enrico Fermi. His physics career came to an abrupt end when, after volunteering to enter Ground Zero of the first atomic bomb test at Alamogordo to collect samples, he was exposed to a half-lethal dose of radiation, meaning that he was banned from further exposure for several years. Being an atomic physicist unable to work with radiation is not conducive to success, so, whilst continuing to teach, he put himself through Harvard Law School (possibly showing a genetic basis for Cliff's own impressive work ethic). Cliff takes up the story: "He had a very nice career because right when he finished law school, the peaceful uses of the atom were being developed, and if you had a reactor or a cyclotron or something and you wanted some patent protection for it, you could go to my Dad, who knew the physics better than the people at your company did, and had a law degree from Harvard, and had joined a very good law firm." From this start, Julius became known as a science attorney, providing legal advice for the first generation of biotech companies and also acting as general counsel for the Salk Institute in San Diego for many years, where amongst other things, he helped Salk scientists patent their discoveries. Cliff again: "When we found Sonic, he basically said: 'You guys need to patent this.' At the time, we had the information that was in those papers—today the thing wouldn't fly, as we had no data that would be of medical use—but at my Dad's suggestion, I talked to Phil and Andy and then we talked to the Harvard Patent Office." As a result, in December 1993, U.S. patent number 5789543, "Vertebrate Embryonic Pattern-Inducing Proteins and Uses Related Thereto," became the first in a series of patents filed by Harvard and the ICRF relating to the Hedgehog pathway.

Someone else at Harvard had woken up to the possibility that developmentally important signalling factors might have relevance to human health, and in 1994, Doug Melton, Cliff's erstwhile postdoc advisor, founded a company, Ontogeny, Inc., with Phil, Andy, Cliff, and Tom Jessell on its scientific advisory board. The Hedgehog pathway patent was an obvious starting point, and following the discovery of the role of Patched and Smoothened in tumours, the early intention to develop drugs to treat degenerative disorders of the nervous system such as Alzheimer's was quickly overtaken by the imperative of an anti-cancer therapeutic. In 2000, Ontogeny, Inc., became Curis, Inc., but the Hedgehog thread remained, culminating in the development of a drug, vismodegib, which bound and inhibited Smoothened, suppressing aberrant Hedgehog pathway activation. After passing very successfully through clinical trials, the Curis drug, developed and commercialised by Genentech and Roche and rechristened Erivedge, was approved for patient use in January 2012. It is still the only treatment for advanced basal cell carcinoma and is a salutary lesson to those who claim that basic, curiosity-driven science has nothing to offer modern medical research.

All good things come to an end, and the three-way collaboration between Cliff, Andy, and Phil did not persist beyond the triumph of their *Cell* papers. Inevitably, trying to avoid stepping on each other's toes whilst simultaneously reaping the rewards of their labours proved too difficult, and the Hedgehog Three went their separate scientific ways. Happily, through the intervening years, during which they have all continued to grow in scientific eminence, the respect that the three have for each other has not diminished, and they all have good memories of their astonishing codiscovery. Perhaps Cliff, always the most loquacious of the collaborators, should have the last word:

> It was a lot of fun. It wasn't that we thought we'd done something incredibly important, it wasn't that we were racing against Jessell and Beachy, who we didn't know about at the time, it was just fun! Phil and Andy

are incredibly smart guys, they're different, they think differently, and it was just fun to play off of both of them. The three of us have remained very good friends over the years, but I sometimes think if we'd carried on working together it would have been incredibly productive as well as enormous fun. All of us have done fine, I'm not complaining, but together we'd have been unstoppable. We could have done whatever we wanted, divided how we wanted—it would have been a really special relationship in developmental biology.

Web Resources

www.nobelprize.org/nobel_prizes/medicine/laureates/1995 Ed Lewis, Janni Nüsslein-Volhard, and Eric Wieschaus on the Nobel Prize website.

www.youtube.com/watch?v=e3HcqGcXls4 Video about Spemann's Organizer.

www.youtube.com/watch?v=Lb6TJzTLg_E *Drosophila* development. Note that the segmentation genes are all active before anything visually interesting is happening (i.e., before 10 h).

Further Reading

Lewis J. 2008. Development of multicellular organisms. In *Molecular biology of the cell*, 5th ed. (ed Alberts B, et al.), Chapter 22. Garland Science, New York.

A clear, beautifully written introduction to developmental biology by Julian Lewis, one of the best in the business.

Slack JMW. 1998. *Egg and ego: An almost true story of life in the biology lab.* Springer-Verlag, New York.

Jonathan Slack's lightly fictionalised account of life in the ICRF labs at Mill Hill, and thereafter.

Quotation Sources

Christine Nüsslein-Volhard-Autobiography. Nobelprize.org. www.nobelprize.org/nobel_prizes/medicine/laureates/1995/nusslein-volhard-autobio.html.

Smith JC. 2000. Not a total waste of time. An interview with John Gurdon. Interview by James C. Smith. *Int J Dev Biol* **44:** 93–99.

Tabin C. 2008. An interview with ... Cliff Tabin. *Nat Rev Genet* **9:** 420.

Glossary

Some of the definitions in this glossary have been modified or taken directly from the glossaries of the following books:

Wolpert L, Beddington R, Jessell T, Lawrence P, Meyerowitz E, Smith J. 2002. *Principles of Development*, 2nd ed. Oxford University Press, New York.

Alberts B, Bray D, Lewis J, Raff M, Roberts K, Watson JD. 1994. *Molecular Biology of the Cell*, 3rd ed. Garland Publishing Inc., New York.

^{32}P: A radioactive isotope of phosphorus. Term used sloppily in molecular biology to indicate any chemical compound containing ^{32}P made for the purposes of labelling biological molecules such as DNA and RNA. For some reason, normally referred to as "P32."

Adenovirus: A medium-sized virus with a linear double-stranded DNA genome. Causes a wide range of human illnesses such as respiratory infections and conjunctivitis.

Allele: One of a set of alternative forms of a gene. In most cells, each gene will have two alleles, each occupying the same position (locus) on homologous (paired) chromosomes.

Amino acid: Organic molecule containing both an amino group and a carboxyl group. The building blocks of proteins.

Amino terminus: The front end of a polypeptide chain or protein.

Anterior: In developmental biology, the head end of an embryo. The antero–posterior axis defines which is the head end and which is the tail end of an animal or structure (in the limb, the thumb is anterior, the little finger is posterior).

Antibody: Protein produced by the body in response to a foreign molecule or organism. In the laboratory, is used to detect and bind tightly to its antigen.

Antigen: Molecule that provokes an immune response, including the production of antibodies.

Antiserum: Blood serum containing a mixture of antibodies.

Apoptosis: Programmed cell death.

ATP: Adenosine 5′ triphosphate, used by many enzymes as an energy source.

ATPase: An enzyme that catalyses the conversion of ATP into adenosine diphosphate, in the process generating energy. ATPases tend to be the power sources driving molecular machinery such as the complexes that replicate DNA.

Autocrine: A form of signalling in which a cell secretes a molecule such as a growth factor that then binds and signals back into the same cell. Tumours use autocrine signalling to force their own growth.

Autosome: Any chromosome except the X or Y sex chromosomes.

Bacteriophage: Tiny virus able to infect bacteria. The foundation upon which molecular biology was built.

Base (A, C, G, T): Used in this book to refer to the purines (A, adenine; G, guanine) and pyrimidines (C, cytosine; T, thymidine; and U, uracil) in DNA and RNA.

Biochemistry: The chemistry of living things.

Blotting: The process by which, subsequent to **gel electrophoresis**, whole gels can be transferred to membranes. The membranes can be probed with labelled DNA or RNA or antibodies, to visualise specific bands. **Southern blotting**: The original gel was loaded with DNA. **Northern blotting**: The gel was loaded with RNA. **Western blotting**: The gel was loaded with protein.

Branch migration: The process in which a crossover point between two DNA double helices slides along the helices.

Buffer: Generic name for the solutions, containing various mixes of chemicals, in which experiments are performed. Called buffers because they contain chemicals that hold the pH of the solution steady.

Carboxyl terminus: The back end of a polypeptide chain or protein.

Cell cycle: The period between the formation of a cell by division of its mother cell and the time when the cell itself divides to form two daughters.

Checkpoint: Point in the cell cycle where progress can be halted until conditions are suitable for the cell to proceed to the next stage. Thus, **checkpoint control**: the mechanism by which the cell detects damage or unfavourable events and halts the cycle.

Cloning: Exact duplication of a molecule or organism. Also used as a verb: "to clone a gene," meaning to produce many copies of a gene by repeated cycles of replication.

Cloning vector: A genetic element, usually a plasmid, bacteriophage, or virus, that is used to carry a fragment of DNA into a recipient cell for the purpose of gene cloning.

Codon: Sequence of three nucleotides in a DNA or messenger RNA molecule that represents the instruction for incorporation of a specific amino acid into a growing polypeptide chain.

Complementary strand: Two strands of DNA are said to be complementary if they can form a perfect base-paired double helix with one another.

Complementation: In genetics, complementation is used to test whether two mutant strains that produce the same phenotype (e.g., a change in wing structure in flies), carry dud copies of the same or different genes. The test only works if the mutation is homozygous recessive; that is, if it only shows up when both copies of the gene are mutated. If, when the strains are crossed with each other, some offspring look normal, then the two mutations must be in different genes; the mutant copy of each gene is being rescued ("complemented") by the normal copy from the other parent.

Conditional mutant: A mutant that only kicks in under certain physiological conditions such as increased temperature.

Cyclins: Proteins that periodically rise and fall in concentration in step with the eukaryotic cell cycle. Cyclins activate crucial protein kinases (called cyclin-dependent protein kinases, or CDKs) and thereby help control progression from one stage of the cell cycle to the next.

Cyclin-dependent protein kinase (CDK): A protein kinase that has to be complexed with a cyclin protein in order to act; different Cdk–cyclin complexes control progression of many of the cellular reactions involved in the regulation and execution of the cell cycle.

Cytoplasm: Contents of a cell that are contained within its outer plasma membrane, but, in the case of eukaryotic cells, outside the nucleus.

Dephosphorylation: The process of stripping phosphate groups off proteins or other biological molecules. Catalysed by enzymes called phosphatases.

Diploid: Containing two sets of homologous chromosomes and hence two copies of each gene.

DNA: Deoxyribonucleic acid. The carrier of genetic information.

DNA sequence: The order of nucleotides in a DNA molecule. Thus, DNA sequencing, the process of determining the order of nucleotides in a DNA molecule.

DNA tumour virus: Double-stranded DNA eukaryotic-specific virus able to cause cancer.

Dorsal: The upper surface, or back, of an embryo. As opposed to the ventral, or under surface. The dorso–ventral axis defines the relation of the back to the under surface of an organism or structure.

Enzyme: Protein that catalyses (accelerates, but is not changed by) a specific chemical reaction.

Eukaryote: Living organism composed of one or more cells with a distinct nucleus and cytoplasm. Includes all forms of life except viruses, bacteria, and archaea.

Excision repair: Process by which DNA is repaired by cutting out the damaged section and filling it in again using the opposite strand of DNA as a template.

Extract (protein, cell-free): What you get if you pop cells open. Depending on the method of preparation, extracts can be full of proteins, DNA, and/or RNA, just the contents of the nucleus, just the contents of the cytoplasm, and so on.

Fibroblast: Common cell type found in connective tissue. Migrates and proliferates readily in wounded tissue and in tissue culture.

Floor plate: Part of the developing neural tube (which goes on to become the spinal cord). Composed of nonneural cells and vitally important for patterning of the ventral part of the neural tube.

G_1 phase: Gap 1 phase of the eukaryotic cell cycle, between the end of cell division and the start of DNA synthesis.

G_2 phase: Gap 2 phase of the eukaryotic cell cycle, between the end of DNA synthesis and the beginning of mitosis.

Gap genes: Zygotic genes (i.e., those present in the fertilised egg, not coming from the mother) coding for transcription factors expressed in early *Drosophila* development that subdivide the embryo into regions along the antero–posterior (i.e., head to tail) axis.

Gel electrophoresis: A method for separation and analysis of DNA, RNA, and proteins, and their fragments, based on their size and charge. Molecules are separated by applying an electric field to move the negatively charged molecules through a gel matrix. Smaller molecules move faster and therefore migrate farther than larger ones.

Gene: Region of DNA that controls a single hereditary characteristic.

Genome: Total genetic information carried by a cell or an organism.

Genotype: Genetic makeup of a cell or an organism.

Germline: The lineage of germ cells, that is, those cells that contribute to the formation of a new generation of organisms; distinct from somatic cells, which form the body and leave no descendants in the next generation.

Gradient (e.g., sucrose, lactose): Running molecules down a gradient is a method of separating DNA, RNA, or proteins by centrifugation, based on their buoyant density. Gradients are made using buffers containing varying amounts of viscous molecules such as sucrose or lactose. The sample is loaded on top of the gradient in a test tube, and the tube is spun at high *g* force over set periods of time. Different-sized molecules migrate down the gradient until they reach the point at which the density of the gradient is the same as their own.

Growth factor: Extracellular polypeptide signalling molecule that stimulates a cell to grow or proliferate. Most growth factors have other actions beside the induction of cell growth or proliferation, depending on the circumstances in which they find themselves.

Holliday junction: A mobile junction between four strands of DNA. The structure is named after Robin Holliday, who proposed it in 1964 to account for a particular type of exchange of genetic information that he observed in yeast known as **homologous recombination**. Holliday junctions are highly conserved structures, from prokaryotes to mammals.

Homeobox: Region of DNA in homeotic genes that encodes a DNA-binding domain called the **homeodomain**. Genes containing this motif are called **homeobox genes**. The homeodomain is present in a large number of **transcription factors** that are important in development.

Homeotic mutation: A mutation that gives rise to a homeotic transformation—the transformation of one structure into another.

Homologous chromosome: One of two copies of a particular chromosome in a diploid cell, each copy being derived from a different parent.

Homologous genes: Genes sharing significant similarity in their nucleotide sequences indicating that they are derived from a common ancestor.

Homologous recombination: The recombination of two DNA molecules at a specific site of sequence similarity.

Homologues: Genes from different species that share a common ancestry or have a common function.

Homology: Morphological or structural similarity due to common ancestry.

Immortalise: To generate a cell line capable of an unlimited number of cell divisions.

Immunoglobulin: An antibody molecule.

Immunoprecipitation: The method by which a specific protein can be fished out of a general mix, using as bait an antibody directed against it.

In vitro: Term used by biochemists to describe a process taking place in an isolated cell-free extract. Also used to describe cells growing in culture as opposed to in an organism.

In vivo: In an intact cell or organism.

Insert (n): A fragment of DNA carried in a cloning vector.

Journal hierarchy: Publishing in certain journals carries a great deal of prestige with it. *Nature* and *Science* are the generalist journals that sit at the top of the heap, with *Cell* not far behind. *Proceedings of the National Academy of Science USA* (*PNAS*) used to be a way of getting interesting data out fast, because Academy members could submit a few papers a year without review. The preeminence of certain journals, and therefore their editors, is a matter of enduring irritation to many scientists, who nevertheless collaborate with a system they dislike in the interests of career advancement.

Junction resolution: Process by which a Holliday junction is cut and resolved into two separate double helices.

Kinase: An enzyme catalysing the addition of a phosphate group to a protein or other molecule. Kinases and phosphatases are the regulatory kings of the cell because phosphorylation can rapidly and dramatically alter the properties of a molecule.

Library: In this book, used to describe a set of fragments of DNA, cloned into an appropriate vector, which together encompass all or a large part of the genes present in a given organism.

Licensing Factor: The name given to the protein factor that determines when cells started replicating their DNA. Biochemically, it is the highly regulated replication complex that binds to origins and initiates replication.

Ligase: An enzyme that joins together two molecules in an energy-dependent process. DNA ligase links two DNA molecules together through a phosphodiester bond. Used in the last step of repair reactions and also as a vital component of DNA cloning, where it is used to glue fragments together.

Linkage: Coinheritance of two genetic loci that lie near each other on the same chromosome; the greater the linkage, the lower the frequency of recombination between the two loci, and the nearer they must be.

Locus: In genetics, the position of a gene on a chromosome. Different alleles of the same gene all occupy the same locus.

Lysis: Rupture of a cell's plasma membrane, leading to the release of cytoplasm and the death of the cell. Many viruses lyse cells as part of their life cycle, and it is also used in the laboratory to get at the cell's contents for a lysate, or extract.

Mass spectrometer: An instrument that can measure the masses and relative concentrations of atoms and molecules. It makes use of the basic magnetic force on a moving charged particle. It can be used as a way of identifying proteins, which all have unique signatures.

Media: Generic name for the solutions needed for growing cells.

Mercaptoethanol: A somewhat toxic chemical smelling strongly of rotten eggs. Beloved of biochemists because it helps stabilise proteins taken out of their natural milieu.

Messenger RNA (mRNA): RNA molecule that specifies the amino acid sequence of a protein. Produced by RNA splicing (in eukaryotes) from a larger RNA molecule made by RNA polymerase as a complementary copy of DNA. It is translated into protein in a process catalysed by ribosomes.

Mitosis: Division of the nucleus of a eukaryotic cell, involving condensation of the DNA into visible chromosomes.

Molecular biology: The branch of biology that deals with the molecular basis of biological activity.

Monoclonal antibody: Antibody secreted by a hybridoma clone. Because each such clone is derived from a single B cell, all of the antibody molecules it makes are identical.

Morphogen: Any substance active in pattern formation whose spatial concentration varies and to which cells respond differently at different threshold concentrations.

Neurulation: The process in vertebrates in which the ectoderm of the future brain and spinal cord—the neural plate—develops folds (neural folds) and forms the neural tube.

Neuron: A nerve cell. The electrically excitable cells of the nervous system that convey information in the form of electrical signals.

Notochord: In vertebrate embryos, a rod-like cellular structure that runs from head to tail and lies centrally beneath the future central nervous system.

Nuclease: An enzyme able to cut nucleic acids, either RNA, in which case they are called RNases, or DNA, in which case they are DNases.

Nucleoprotein: Any protein that is structurally associated with DNA or RNA.

Nucleosome: Structural, beadlike unit of a eukaryotic chromosome composed of a short length of DNA wrapped around a core of histone proteins; the fundamental subunit of chromatin.

Nucleoside: Compound composed of a purine or pyrimidine base linked to either a ribose or a deoxyribose sugar (in DNA, the sugar is a deoxyribose, and in RNA, it is a ribose).

Nucleotide: Nucleoside with one or more phosphate groups joined to its sugar moiety. DNA and RNA are polymers of nucleotides.

Nucleus: The central compartment of a eukaryotic cell, containing DNA organised into chromosomes.

O levels: Between 1950 and 1988, the exams taken by 15–16 year olds in Britain. Typically sat in eight to 12 subjects. Followed by a smaller number of A levels, taken at 17–18 years of age, and necessary for university entrance.

Oligonucleotide: Polymerised nucleotides. Generally used to describe short pieces of DNA.

Oncogene: Gene capable of causing cancer. Typically, a mutant form of a normal gene (proto-oncogene) involved in the control of cell growth or division.

Oncoprotein: Protein product of an oncogene.

Organiser: In developmental biology, a signalling centre that directs the development of a whole embryo or of part of the embryo, such as a limb.

Origin of replication: Specific region in a DNA molecule from which replication starts. In bacteria, there is just one origin of replication. In mammals, there are tens of thousands.

Pair-rule gene: *Drosophila* gene involved in specifying the boundaries of parasegments. They are expressed in transverse stripes in the blastoderm, each pair-rule gene being expressed in alternate parasegments.

Parasegments: In the developing *Drosophila* embryo, parasegments are independent developmental units that give rise to the segments of the larva and adult.

Peptide: Very short linear polymer of amino acids.

Phage: See bacteriophage.

Phenotype: The observable or measurable characters and features of a cell or organism.

Phosphorylation: Process of adding phosphate groups onto another molecule. Accomplished by kinases.

Plasmid: A small circular DNA molecule that replicates independently of the bacterial genome. Used by molecular biologists for cloning genes.

Platelet: Cell fragment, lacking a nucleus, that breaks off from a megakaryocyte in the bone marrow and is found in large numbers in the bloodstream. It helps initiate blood clotting when blood vessels are injured.

Polymerase: Enzyme capable of synthesising polymers of nucleic acids.

Polyoma: Small, double-stranded DNA tumour virus.

Polypeptide: Linear polymer composed of multiple amino acids. Proteins are large polypeptides, and the two terms can be used interchangeably.

Positional cloning: The cloning or identification of a gene for a particular disease based on its location in the genome.

Positional information: Can take the form of, for example, a gradient of an extracellular signalling molecule such as Sonic hedgehog, which tells the cells where they are in a developing structure and instructs them to develop into the right cell type for their location.

Posttranslational modifications: Modifications made to proteins after they've been made; includes actions such as phosphorylation and ubiquitination.

Postdoc: Someone at the post-PhD (i.e., postdoctoral) stage of their scientific career. Also used as a job definition and a verb. Postdocs are on short-term contracts, typically for 2 or 3 years; postdoc contracts are renewable but never permanent.

Posterior: In developmental biology, the tail end of an embryo. The antero–posterior axis defines which is the head end and which is the tail end of an animal or structure (in the limb, the thumb is anterior, the little finger is posterior).

Primary cells: Cells taken straight out of an organism into tissue culture. Will grow in tissue culture for only a few generations, or not at all, if you're unlucky.

Prokaryote: Organism made of simple cells that lack a well-defined membrane-enclosed nucleus; bacteria or archaea.

Protease: An enzyme that can digest proteins.

Protein: The major macromolecule constituent of cells. A linear polymer of amino acids linked together by peptide bonds in a specific sequence.

Proto-oncogene: Normal gene, usually concerned with control of cell proliferation, which when mutated into an **oncogene** can cause cancer.

Pseudo-autosomal: The regions on the X and Y sex chromosomes that are similar enough to be able to pair with each other.

Ras: One of a large family of GTP-binding proteins that help relay signals from the cell-surface receptors into the nucleus. First identified in several mutant forms as the oncogene of several different rat sarcoma viruses.

Reading frame: The phase in which nucleotides are read in sets of three to encode a protein; an mRNA molecule can be read in any one of three reading frames.

Receptor: Molecule able to recognise and bind to a specific extracellular signalling molecule (ligand) to initiate a response in the cell. Cell-surface receptors, rather like Winnie the Pooh after too much honey, are stuck in the plasma membrane with their ligand-binding domains exposed to the outside of the cell and their signalling domains hanging in the cytoplasm. Intracellular receptors also exist, which bind the ligand once it has diffused through the plasma membrane into the cell.

Recombinant DNA: Any DNA molecule formed by joining DNA segments from different sources. Recombinant DNAs are widely used in the cloning of genes, in the genetic modification of organisms, and in molecular biology in general.

Recombination: Process by which chromosomes or DNA molecules are broken and the fragments rejoined in new combinations. Can occur in the living cell or in the test tube using purified DNA and enzymes that break and religate DNA strands.

Replication: Process by which a cell makes a new copy of its genome, before cell division.

Resolvase: The enzyme that cuts Holliday junctions, resolving the four-way structure into two separate double helices.

Restriction enzyme: One of a large number of nucleases that can cleave a DNA molecule at any site where a specific short sequence of nucleotides occurs. The workhorse enzymes of molecular cloning and analysis of DNA.

Restriction map: Diagrammatic representation of a DNA molecule, indicating the sites where various restriction enzymes can cleave.

Restriction site: Each restriction enzyme has a specific sequence it likes to cut, known as its restriction site.

Retrovirus: RNA-containing virus that replicates in a cell by first making a double-stranded DNA intermediate.

Reverse genetics: In reverse genetics, the functional study of a gene starts with the gene sequence rather than a mutant phenotype. Using various techniques, a gene's function is altered, and the effect on the development or behaviour of the organism is then analysed.

Reverse transcriptase: Enzyme, present in retroviruses, that makes a double-stranded DNA copy from a single-stranded RNA template. The only transgressor of Crick's Central Dogma that DNA makes RNA makes protein.

RNA: Ribonucleic acid. Polymer formed from ribonucleotide monomers.

RNA splicing: Process in which intron sequences are excised from RNA molecules in the nucleus during formation of messenger RNA.

RNA tumour viruses: A subset of viruses of the retrovirus family, able to cause tumours in the animals they infect. Source of retroviral oncogenes.

S phase: Period of the eukaryotic cell cycle during which DNA is replicated.

Segment polarity gene: In *Drosophila*, involved in patterning the parasegments and segments.

Segmentation: The division of the body of an organism into a series of morphologically similar units or segments.

Sequencing: Process of determining the linear sequence of DNA, RNA, or protein.

Serum: What is left from blood after all of the blood cells and clotting factors are removed. Serum includes all proteins not used in blood clotting and all the electrolytes, antibodies, antigens, hormones, and growth factors. It is a rich source of nutrition for tissue culture cells.

Signal transduction: The process by which a cell converts an extracellular signal into a series of intracellular responses, often leading to changes in gene expression.

Sister-chromatid cohesion: A chromatid is one copy of a replicated chromosome that is still joined at the centromere to the other copy, its sister chromatid. **Cohesion** refers to the process by which they are joined. Sister chromatids are separated at mitosis, when they are allotted to the new daughter cells.

Somatic cell: Any cell other than a germ cell. In most animals, the somatic cells are diploid.

Somatic mutation: Genetic alteration acquired by a cell that can be passed to the progeny of the mutated cell in the course of cell division. Working out how to engineer somatic mutations was the breakthrough that started the human genetics revolution, amongst other things.

Southern blotting: Technique in which DNA fragments, separated by electrophoresis, are immobilised on a membrane. Specific molecules are then detected with a labelled nucleic acid probe. Named after its inventor, Ed Southern.

Strand exchange: Process by which homologous recombination is initiated by RecA in bacteria and Rad51 in eukaryotes. A single invading DNA strand extends its partial pairing with its complementary strand and displaces the resident strand from a DNA double helix.

Substrate: The molecule on which an enzyme acts.

Supernatant: The liquid lying above a solid pellet following centrifugation.

SV40: Small double-stranded DNA tumour virus.

Sympathetic ganglia: A ganglion is a biological tissue mass, most commonly a mass of nerve cell bodies. Sympathetic ganglia are therefore ganglia of the sympathetic nervous system. They deliver information to the body about stress and impending danger and are responsible for the fight-or-flight response. They contain approximately 20,000–30,000 nerve cell bodies and are located close to and on either side of the spinal cord in long chains.

Syncytiotrophoblast: The epithelial covering of the embryonic placental villi, which invades the wall of the uterus to establish nutrient circulation between the embryo and the mother.

TDF (testis determining factor): The gene responsible for maleness. Shown by Peter Goodfellow and Robin Lovell-Badge's laboratories to be *SRY*.

Temperature-sensitive (ts) mutant: Organism or cell carrying a genetically altered protein or RNA molecule that performs normally at one temperature but is abnormal at another (usually higher) temperature.

Tissue culture: The process of growing cells or tissues outside the body. Generally refers to animal cells.

Transcription: Copying of one strand of DNA into a complementary RNA sequence by the enzyme RNA polymerase.

Transcription factor: Loose term applied to any protein required to initiate or regulate transcription. Includes both regulatory proteins as well as proteins involved in the actual transcription machinery. In terms of the regulatory proteins, transcriptional activators switch on transcription, and transcriptional repressors inhibit it.

Transfection: Technique for introducing foreign DNA into mammalian and other animal cells. The introduced DNA is sometimes incorporated permanently into the host cell's DNA.

Transfer RNA (tRNA): Set of small RNA molecules used in protein synthesis as adaptors between mRNA and amino acids. Each type of tRNA molecule is responsible for bringing one particular amino acid to be incorporated into a growing peptide chain.

Transformation: Heritable alteration in the properties of a eukaryotic cell, usually referring to the acquisition of cancer-like properties following treatment with a virus or carcinogen, or expression of oncogenes. The next step beyond immortalisation. Transformed cell lines are used to study cancer in vitro.

Transformation (yeast): Introduction of a foreign DNA molecule into a yeast cell; usually followed by expression of one or more genes from the newly introduced DNA.

Transgenic: An animal (or plant) that has stably incorporated DNA from another organism and can pass it on to successive generations. The DNA may encode a gene, engineered to be expressed in the transgenic animal and therefore called a **transgene**.

Translation: Process by which the sequence of nucleotides in a messenger RNA molecule directs the incorporation of amino acids into protein; occurs on a ribosome.

Tritium: A radioactive isotope of hydrogen. Also written as ^{3}H, but always called tritium.

Tumour-suppressor genes: Genes whose function is to protect the cell from cancer. If both copies of such genes are inactivated, the cell is at far greater risk of becoming cancerous. Also known as **anti-oncogenes.** *p53*, *RB*, *BRCA1*, and *BRCA2* belong to this class of gene.

Tyrosine kinase: Enzyme that specifically phosphorylates the amino acid tyrosine in proteins. Extremely important for regulation of a multitude of cellular processes, including signal transduction.

Ubiquitin: A small protein that is attached to other proteins as a post-translational modification. Ubiquitination is a sort of address label for proteins. Depending on how many ubiquitin molecules are bound, and where, ubiquitinated proteins are sorted into different pathways, which can move them into, out of, or around the cell, or send them to be scrapped.

Vector: In molecular biology, an agent (virus or plasmid) used to transmi genetic material into a cell or organism. Used to clone DNA fragments.

Ventral: In developmental biology, the underside of an embryo or animal. The dorso–ventral axis defines the relation of the back to the front of an organism or structure.

Zone of polarising activity: The polarising region at the posterior margin of the limb bud. Responsible for producing the signals that specify **positional information** for the antero–posterior (thumb to little finger) axis of the limb.

Index

Page numbers followed by a *p* indicate a photo.

D

P